The Dance of the PIG

Book Four - Exile

A Cosmology and World History
Based on the Cabala
Spanning the Last 266,455,777 Years

dovid krafchow

ISBN 979-8-2408-2682-5 (paperback)
ISBN 979-8-2408-2683-2 (digital)

dovidHouse Publishing
Yosef Chaim #2
Nachlot, Jerusalem
Israel

dovidHouse.com

dovidHouse
PUBLISHING

Contents

Introduction

Victory, represented in the Six Days of Creation as the fourth day, when organic life takes hold and the angels are created—designated by the word Netzach, the 7th of the Ten Sefirot; during the fourth day of creation the planets are arranged in exacting orbits regulated to future events destined for the Earth. The fourth day of creation represents the thousand years of human history upon planet Earth when animated life takes root in the world.

In the fourth day of creation quickly replicating life abounded in the heavens or in the water, but the Earth was yet barren. These primal life creations, in two similar ethers of ocean and sky, have in common the necessity to be submerged and immersed within their source: like fish in the sea, birds in the sky and the planets in their orbits. The energy of Netzach is reflected in defining elements into a logic-based form. Chaos preceded form, but only in an infinite arena where time had no end could such a postulation be considered. The Torah teaches, Life on Earth has a fixed time of six thousand years plus a thousand years of Woman and Peace..

The Ten **Seferiot/Lumunaries** reflect the nature of confine—we can never get beyond ten, yet the cabalists do go beyond ten through the apparatus of **Netzach/Victory**, representing the inner storehouse of the king's treasures—in order to attain victory, the king will hold nothing back. The necessity of the king to retain sovereignty over the people is endemic to the definition of king—no king without a people; few kings in the world have been deposed; to maintain a state of victory and celebration is incumbent upon the king.

The king needs to be in a continual state of victory and is willing to give everything for that end; the king is the lowest level of Heaven directly connected to the Earth. Every human being is imbued with this aspect of victory—to win at any cost, not to severe our innate connection with creation. The term Netzach can also mean: Forever. Essence stands beyond time and place and therefore **Nezack/Victory** is related to the concept of infinity. The job of the Jewish People is to bring that essential infinity beyond ten levels down to the Earth where the confines of our world could merge with the undefined endless possibilities offered by corporeal life for six thousand years.

To accomplish this task the Jewish People, each one an incarnation of the soul existing like a candle within the sun, prior to creation, were given the Torah to bring and share on the Earth. The Torah, like the soul of the Jewish People, also existed prior to creation; God gazed upon the Torah and from the reflection came the intricate code guiding creation. God promised the soul: If you go down into creation, become a people and accept upon yourselves the obligation of the Torah, then I will bring Moshiach before the year 6000—we are quickly approaching that end.

The first two thousand years of human creation upon planet Earth were devoted to the previous world called **Tohu/Chaos** whose physical manifestation was on Mars and whose life forms were conveyed through the median of water down to the Earth where life slept in the ground waiting to be awakened. Called **Shaverot HaKalim/ Breaking of the Vessels** the fecund organic source was delivered to the ground and dispersed throughout, during the first two days of creation. Similarly, the first two thousand years from first man until the advent of Avraham was a turmoil of life. The first two thousand years of creation were a chaotic amalgam of light and darkness, the primordial soup of creation.

The kindness of creation was reflected in the ethereal human beings living a thousand years with knowledge and ability enough to transport throughout the Solar System finding a home on various worlds of different ecological systems from the Earth until Pluto. However, the final resting place for the human being would be on the Earth, which had been flooded from the heavens 1666 years into creation of the human being on this planet. All that happened was indicative of the first two days of creation.

The last part of the two thousand years of chaos was taken up by the Tower of Babel Where Avraham lived and eventually left in the year 2000 at the age of 52—Avraham had inherited the soul likened to a candle in the sun. At the age of 75 Avraham descends to the Earth, taking on a body of *dust and ashes*. At a hundred years old Avraham and Sara have a child— Yitzchok, to whom Avraham taught Hebrew as a mother language.

Avraham also had a child, Ishmael, from the Egyptian princess Hagar; She is the father of the Arab People and to whom Avraham taught Arabic, another mother language. Yitzchok had twins: Yaakov progenitor of the Jewish People through the Twelve Tribes of Yisroel and Esau father of Rome. These three peoples: Arab, Jewish and Roman were to be the crucible in which the human spirit would be thrown to remove impurities. From this cauldron of boiling Roman anger, wild Arab behavior and Jewish stubbornness the rest of the world were completely unprepared.

After the flood, the world grew a much more moderate human being, but without the influence of these peoples, the coming of Moshiach could not be accomplished. Moshiach was being transferred through the Jewish People. The Jewish People would have to undergo a lot of life, but in the end the reward will be to succeed in mothering Moshiach into the world. The languages of Arabic and Hebrew are spiritual language able to support prophesy; the undisputed original text of the Torah and the Koran remain unchanged since received.

The third millennium, the beginning of the two thousand years of Torah, was the culmination of the seed planted by Avraham and Sara. Through the distillation of a new language in a world stricken dumb from the crash of the Tower of Babel, the conquest of ideas conveyed by advanced languages overcame the brute strength of empires. The first phonetic languages Arabic, Hebrew and the Aryan/Sanskrit taught by Avraham to his six children of his old age (Avraham lived to the age of 175) quickly overcame the stilted and cumbersome language contrived by man called Cuneiform.

The Arab People merged with their neighbors; through the strength of their language Arabic and their connection to Avraham, they dominated and absorbed. The Jewish People, on the contrary, were extremely careful about their genealogy—with the notable exception of figures like Tamar, and Yitro who were not descended directly from Avraham but imbued with the nature of the Jewish People with an uniqueness elevating the entire nation to a higher level than they could have accomplished by themselves. During the fourth day of creation the stars became fixed in their place; so too, the Jewish People became a nation beginning a dynasty of kings extending until the conclusion the the sixth day, now.

During the thousand years of **Netzach/Victory,** the fourth millennium of human history, the Jewish Peoples gave rise to kings—some great and some wicked—while Rome waited and incubated. The Arab People did not have a text, but nonetheless established their people through their language taking hold in Saudi Arabia; the Jewish People expanded the Five Books of the Written Torah into twenty four books, chronicling the prophets and the writings, during the first half of the fourth millennium. The constant growth of the Hebrew language through the twenty- four books added complexity and nuance to the already obscure short teachings handed down by Moshe— Five Books of the Written Torah.

Much of the Oral Tradition received during the forty years wandering in the Wilderness had been forgotten over the centuries; nonetheless, the 600,000 words inscribed into the Written Torah include the three levels of Oral Tradition invisibly embedded within the text, in the form of pronunciation, the inflection and a differentiation in various crowned letters without any seemingly reasonable explanation. The way in which the Jewish People retrieved these lost ideas and hidden knowledge was through the constant argumentation required to clarify minute details.

Immediately after the Five Books of Moshe was completed, the next book began—the story of the conquest of the land and the settling of the Jewish People. God's promise, to bring these people out of bondage and return them to the land promised to Avraham, had been kept, despite the overwhelming odds. Through the miraculous feats occurring to the Jewish People, the world took notice. Each additional book brought new ideas, miraculous stories and visions of the future The diversity with which the Jewish People kept the commandments, learned the Torah and increased the law furthered the story; out of the story emerged the metaphor portraying the hidden the secret.

Each subsequent generation added their stamp to the Torah in a non-contradictory fashion expediting endless complexity throughout the Torah. This malleable oral knowledge was the safeguard against the Torah becoming dogmatic. There is no term in Hebrew for word Religion nor was dogma practiced by the people; and though the tribe of Levi officiated within the **Beit HaMikdesh/Temple**—the land, the buildings and the sacrifices belonged to all the people. God was One and the Jewish People were One; the Shakina dwelt between the staves on each side of the Ten Commandment hewed out of a sapphire cube.

The **Beit HaMikdesh/Temple** was the crowning achievement of the Jewish People, the final resting place for the Ten

Commandments; the Jewish People had become a righteous nation as God had intended and now warranted the presence of God in a revealed manner so all the world could see, understand and from this example find ways to bring God down to Earth. In the fall, seventy bulls would be sacrificed for the seventy nations of the world; emissaries would be sent to participate in the holiday of **Succot/Covering** from throughout the world.

The tribe of Zebulon was greatly praised, since living by the ocean they naturally became sea-faring merchants who through their commerce induced people of other nations to come to the Land of Yisroel where they would see one people, eating one food and praying to one God. Peoples of the world were induced to give up the Seven Commandments of Noah and instead obliged themselves to perform the 613 **Mitzvot/Commandments** of the Torah thereby entering the ran... ded Dovid, ...

...ncept of king which ...ed **Malchut/Royalty**. ...hepherded the Jewish ...epherds correspond to ...ower seven of the Ten ...ets from Earth to Pluto. ...ing 6000 when we cul- ...the advent of the final ...th called The Thousand

...d an important ... to the Torah. ... Torah have to ...ing the poor ...ere admoni- ... income so ...ntered into ... land and

...ham had ... brought

the Torah from the heavens down to the Earth and Dovid had become the first in the line of kings who would culminate with Moshiach—bringer of peace to the world. Before Moshiach could come the Temple had to be built and though Dovid wanted to build the Temple saying, *Why should I live in a house when the Law of God remains in a tent?* God did not permit Dovid to build the Temple, saying *Your hands are too full of blood.*

The Talmud-Book of Law notes: though Dovid's hands were full of blood from war, he had found use for this knowledge by inspecting the embryos of miscarriage for the sake of making peace between man and woman. Dovid could tell the sex of the intended baby which meant he could permit the wife to return to her husband at an early date; according to the Torah: women were required to separate from her husband after birth for 33 days if a boy or 66 days if a girl. The couple wanting to have sex again ne King of Yisroel, to determine the sex to be male.

Dovid is the exemplification of the cc
is related to the tenth of the Ten Sefirot cal
Dovid is the last of the seven leaders who
People into pleasant pastures; these Seven S
the seven days of the week, related to the
Sefirot as represented by the seven outer plan
The seven thousand years of creation is nea
minate with the coming of Moshiach and
millennium of human history upon the Ea
Years of Woman and of Peace.

The cabalists point out, the first letter of Avraham, Moshe and Dovid make the word Adom—first man. Through the work of the Jewish People the ancestry of Adom will find their completion as the Midrash-Book of Metaphor says, If the nations of the world recognized the great value of the Jewish People they would surround the Land of Yisroel with armies so no one would hurt them.

Chapter One

Question of Zero

NOTHING

Existence fears zero, existence exists relying on endless divisions divided endlessly producing infinite ones, never reaching zero. This common fear of zero is found in the concept of creation coming out of nothing exemplified by zero, eventually turning back into the nothing of zero leaving in the wake of the vast emptiness of nothing leaves the question, why?

Nothing is really something—in fact, nothing is the first thing created by the Creator. Nothing came from the infinite possibilities within the Creator's ability to manifest, called by the cabalists, **OreAinSof/LightWithoutEnd**. The term **Ore/Light** is used because of illuminating nature of light, but the manifestation of God's endless potential was far beyond light. The manifestation of nothing out of the light allowed for revelation to happen within the darkness of nothing; where possibilities to aggregate differentials by allocating discrepancies and assigning levels within the bubble of nothing.

Out of the potential endless possibilities surface, a structured creation emerges as insignificant as a bubble on the water. If all of creation was condensed into a ball and if that ball was thrown into the ocean of God's endless light, creation would not comprise even one drop in that ocean of light. Creation, as we know it, is the manifestation within the tiny bubble of nothing amid the endless potential of something.

Once created, nothing began to expand and would have consumed all the potential if God had not *shouted* at creation to stop by saying **Di/Enough**. The alphanumerical equivalent of **Shadai/ The One Who Said Enough** is 314 the first three numbers of pi. Equating creation with time and space is a result of the mathematical perfection represented by the infinite expanse of pi inhabiting the bubble of nothing, so something can exist. A result of being imbued with the endless characteristics of pi was reflected in the birth of dimension. God wanted to be revealed in the insignificance of nothing and because of God's desire, creation was created in dimension after dimension until this physical world—the end of creation.

The concept of king holds the secret to the creation of nothing. The king is the intermediary between Heaven and Earth on various levels and degrees. On the most mundane level the king intercedes between the Creator and the people and on the more sublime level, the concept of king separates Creator from creation. The king rules from a distance and there is no greater distance than zero. Zero to One is a greater distance than one to infinity because infinity is made of endless ones whereas the distance bridging zero to one is beyond all calculations.

Though the nature of **Malchut/Royalty** is patently female. Royalty is represented by the male king; as portrayed in the Ten Sefirot, Malchut is the only Sefira with curves, turning while the other nine lines only point. The only two geometric figures in creation are the line and the curve; the curve is made up of infinite lines called vectors and each line extended to infinity will return to itself in the form of a circle. The celestial dance between the curve and the line conducted by the rule of pi.

The rotating nature of **Malchut/Royalty**, like a revolving door, shields creation from the immense and overwhelming energy from the Creator filtering down through the narrowing circles of dimension. Malchut opens her Royal mouth and pronounces creation—her words are eternal; likened to the lips from which creation proceeds, Malchut is uncannily like speech—*God spoke and the world was.* To speak one must have nothing in their mouth, then out of nothing speech suddenly occurs; the ability to speak is shared only between the Creator of creation and the created human being.

The paradigm of man, woman and child as implied from the very beginning of creation through the implicit separation between Shechina (Malchut) and the other nine Sefirot leading to the subsequent birth of child (creation) replicated endlessly everywhere throughout the confines of nothing; wherever there is nothing creation pervades. The circle of nothing which began creation as Malchut, the "Mouth of God" ultimately concluding creation in the form of a circle; the line, the other nine of the Ten Sefirot, entered into the circle and thereby are brought into life.

This dance of creation portrayed through the curve and the line is a primordial message imbued with infinite wisdom from the Creator. The production of nothing left bits of debris each set in an orbit with a destination. In the nothing of creation, existence, away from the overwhelming forces of the **OreAinSof/LightWithoutEnd**, became something. And that is how something came out of nothing.

HUMILITY

At first, creation was bashful and unwilling to be detached from the Creator, but finally opening an eye to consciousness, creation had to admit a separation had been forged; the Oneness had been divided into a world of divisions that replicated endless degrees of creation until the Creator yelled, **Di/Enough.**

The Zohar-Book of Secret begins from the point of nothingness when space expanded at infinite speed until the Creator said, **Di/Enough**. The Creator then became known to creation as Shadi: the One who said Enough. Implicit in **Di/Enough** is for formula for physical creation—pi,

3.14 has an infinite number of digits that do not repeat, constantly reconciling between the line and the circle. Scientists are surprised by the unlikely places in which they find Pi, until forced to admit the physical universe has a symmetrical geometry; perhaps, not until Einstein in 1906 of the Roman calendar and 5666 in the

Hebrew calendar did the notion of symmetry take on metaphoric meaning.

Light was both particle and wave not dissimilar to life which is made of body and soul. Since the year 5666 or 1906 the impenetrable problem of atomic proportions has been revealed for good and bad. Atomic reaction became synonymous with the destruction of the world, while the more ephemeral aspects of nuclear knowledge deriving from additional metaphoric information derived from the dismantling of the atom has not been adequately explored nor has this knowledge been fully acknowledged for its profundity—like the three triangular forms of quarks from which all nuclei is formed.

The concept of **Malchut/Royalty**, the connector between zero and one is exemplified by the atom which in turn represents the attribute of humility, synonymous with nothing, making the king the ultimate arbitrator—serving both God and the people through humility. Because of humility, the ego appears less important than the soul. It is through this disparity of ego combined with humility, freedom of choice is forged. Humility makes place for the Creator to interact within creation–in this way, God is known in low.

The purpose of separating creation from the Creator is to allow a dialogue to begin; in the same way oneness became plurality so too did monologue of creation turn into dialogue with the Creator. The reason why **Malchut/Royalty** is exemplified by speech is because a vocal statement declares individuality king; or in more parochial terms: our opinion. The apparatus of speech and the mechanism of the body providing the input depicts the constant process of creation coursing through everywhere; ten is endemic and ten is Dovid, the Seventh Shepherd, the tenth sefera.

The lower seven **Sefirot/Lumunaries** fell from the previous world, the remnants of the Seven Kings of Tohu who lived and died. Because each of the Seven Kings of Tohu wanted everything their world broke, the sparks which fell when the vessels broke became embedded into the physicality of our world; the spark within physicality draws terrestrial life to want more, to live and want more—the will to live. The imperfection of life on Earth provides needed humility to the sparks from Tohu, as reflected in our struggle to communicate with words.

Speech is peculiar to the human being, an innate ability to reveal thoughts; speech is the most hidden of the intrinsic human powers; through speech the human being is able to conquer the animal and rule the world. Through the speech of Dovid, first king of the Jewish People, the world would hear the praises of God for over three thousand years. Dovid Speak—the words of **Tehilim/Psalms**—are short phrases of emphatic visions and endless illusions to secret ideas imbedded throughout the Torah.

In the Book of Shmuel, Dovid is the main character, but the Book of Tehilim is filled with Dovid's chaotic meanderings—so peppered was his life. It was Dovid who said, *My mother and father have abandoned me but God will take me in.* Dovid knew being alone brought him closer to the Creator; he also knew the constant strife in his life was a test. Dovid passed so many tests that eventually he cried out to God, *Test me, I am pure!* And God did.

BATSHEVA

Dovid King of Yisroel whose authority spread from Yehuda until Daun, saw a beautiful woman standing on the roof tops bathing and he wanted her. Dovid sent messengers to ascertain who the woman was and found out she was the wife of a soldier who was presently deployed in battle. It was customary for the Jewish People when a man had to go to war or embark on a long journey he would write his wife a paper of divorce, incase the man disappears the woman would be free to marry another.

Therefore, according to the law, Dovid was allowed to have her. When the husband returned home and realized that his wife had been violated by the king he refused to enter into his house. Seeing the man was not worthy of the woman the king sent him back into battle where he died and the king claimed the wife for his own.

Dovid failed the test, seemingly because he could not stand up to the temptations of flesh; Dovid, who had said hearing how his friend Yonaton was dead, *I loved him more than woman*, was unable to stand up to the woman? Her name was **BatSheva/Daughter of Seven**, gave birth to Shlomo who would retain the crown after his father, Dovid passed away; Shlomo built the great Temple in the place his father had purchased and arranged the site—the two sides of Malchut, Dovid and Shlomo came together in **Yerushaliam/Awe** at the Temple Mount set between two hills replicating the original **Malchut/Royalty** from whence the world came into existence.

The Talmud-Book of Law maintains, Dovid was not wrong in taking the man's wife as was the king's privilege to take her while she was divorced; since Dovid saw through prophecy she had the seed of Moshiach, therefore he wanted her—Dovid's misdeed had been, he wanted her too much. Even before the messengers arrived to tell the king, BatSheva's husband was dead, Dovid had already taken her; he saw in prophecy, the man was dead, but he could not wait for the messengers to actually tell him, as obliged by the Rabbis.

Dovid had to admit he was not perfect; although he strove to be perfect and begged God to test him, yet he failed the test—he prevented the heaven coming to Earth. Using prophesy he went out to meet the news before the news became apparent. The imperfection of Dovid caused him to strive beyond what he actually was capable of doing; to trust God would help him as Dovid said, *Even as the sword is upon your neck, cry out to God for help.*

Dovid was not like other kings, as he reminded God, *Other kings awake in the third hour of the morning, while I awake at midnight to praise your Name.* And true it was, Dovid would place his harp in the opening of the northern window because, until this day, a wind comes up in north precisely at midnight; his harp would quiver; he would arise and sing, *Wake up my harp and I will wake the dawn.* The Talmud-Book of Law says, actually Dovid never slept, he just nodded off between evening and midnight then he would spend the night in study before attending to the business of the nation.

Dovid was beyond politics whatever people thought of him, he was there to serve and love the people and in that way he would serve and love God.

THE PLACE

Other than bringing the seed of Moshiach closer to the Earth, Dovid also found and purchased the place where the **Beit HaMikdesh/ Temple** would be built. First he procured it with money and then he prepared it for the building, but God would not let him continue saying, his hands were too full of blood to toil in a building made to bring peace to the world—that would be for his son Shlomo. But Dovid would prepare the place.

The Place was situated between two hills, mounds on each side of a valley; it was there the two brothers had kissed—each one wanting to give the riches of their lives to the other. It was in this place the first human being had been gathered together from the inanimate bits of matter spanning the outer seven planets of our solar system. Creation had begun with the engraving of place into the hard light until Earth became the intention of the Creator at the end of creation where God could have a dwelling place in low.

This was the place where Avraham went to slay his son Yitzchok; the place where Yitzchok was praying when Eliezer servant of Avraham brought Rivka to fill the place of his mother Sara who had recently died–the same place where Yaakov slept and saw in a dream a ladder going into heaven. Upon the ladder were angels going up and angels going down. There is much discussion in each generation as to why the angels are first going up then going down, since angels originate in Heaven then come down to the Earth.

In the Midrash-Book of Metaphor explains, the Beit HaMikdesh was situated between the mounds because this is the most beautiful place of the Ox—between the shoulder blades; but, the cabalists see it in a very different orientation: the two mounds like thighs and the place of the Bait HaMikdesh the sex of woman. There the animal was sacrificed while the people went in and out; there was a place of great and pervading light and incense so sweet one could hardly stand; there was the place of the two cubes of the Ten Commandments progenitor of all words in the Torah, hidden in the Holy of Holies like the two ovaries of the woman.

In cabalistic and talmudic lingo, the sex of the woman is known as **HaMakom/The Place**. In the Zohar-Book of Secret says, The seven land masses of the world are arranged in the form of a human body; the sex of our female world is in the Mediterranean Sea that ends at the shores of **Eretz Yisroel/Land of Yisroel**—the place of conception is the **Beit HaMikdesh/ Temple**. Avraham had called this Place a mountain; Yitzchok had said this Place was a field, but Yaakov called it a house—a holy house, a place full of awe.

Yerushaliam-Yira Sholom/Complete Fear is understood as Awe. This is the Place where Moshiach will come and the world will be transformed into peace. The fourteen generations since Avraham had led to the fifteenth—**Shlomo/Complete** like the full moon of the Jewish People— radiating a great and gentle light into the darkness of humanity. The **Beit HaMikdesh/Temple** would last for four hundred and ten years before being destroyed by Babylon. There are hints hidden in the Torah and elucidated throughout the writings of the prophets of three Beit **HaMikdeshim/Temples** that would be built before the year 6,000—the present date of human history is 5777.

The Place of the **Beit HaMikdesh/Temple** and its construction and configuration was an encapsulated philosophy with many particulars. The Egyptian legacy of the pyramids utilizing magic to hide pi could not comprehend the meaning of infinity and balked at the idea of void. They lived in numbers that attenuated form describing creation in human terms of endless ones—God was left out of the equation, zero to infinity were denied, reduced to just a big bang. But, the Place of the **Beit HaMikdesh/Temple** was a stairway into Heaven for all to come and be awe-inspired.

In Egypt, the Pharaoh was the ambassador to the spirit world; the people served the Pharaoh and the Pharaoh served the gods—an internal operation. The Creator was left out through the disregard of zero and infinity. Zero recognizes the world came out of nothing while infinity acknowledges forever. The Torah and the Temple embrace and merge with the nothing; the Arab People brought zero into mathematics. To the Creator, zero and infinity are the outermost limits of the created world contracted down into a ball which when thrown into the Ocean of God would not comprise a drop.

Zero and infinity are the gateways through which God and humanity speak—both incomprehensible to the human mind. The only human experience related to the illogical phenomenon of zero and infinity is sexuality. Crying out to god during sex is not uncommon. Whereas, others of those times incorporated sex into their ritual, the Jewish People celebrated sex in the metaphor built into stone where miracles abounded due to the gust of spiritual wind blowing through the portals from Heaven: pillars of smoke would stand erect while metal angelic figures would embrace; there were no flies, no bugs, no snakes or scorpions.

The Jewish People celebrated zero and infinity in the personification of the Ten Commandments engraved with the *Finger of God* through a cube of sapphire displaying a different text on each side. The Ten Commandments were situated in a room with the measurements of twenty yet from each side of the Ark that encased the Ten Commandments measured ten, since the ark existed beyond space. The two staves fasten to the side of the Ark pressed out in the form of two breasts against a woven curtain portraying two different animals on each side with the same weave.

The **Kodesh HaKodeshim/Holies of Holies** was the fulcrum point of the **Beit HaMikdesh/Temple** which was configured in the form of the human being with its divisions and corridors leading to various chambers and hidden tunnels. On the holidays when the people filled the vast halls of the Temple until there was no room to bow down but, miraculously when the Name was spoken by the Kohan Gadol, the head official, all the people would prostrate on the ground—where there had been no room suddenly there was room.

PEACE

Dovid had secured th erected the house. Dovid was asked to be king by the people because of the great strength of his convictions—Dovid exemplified being Yisroel. He was from the tribe of **Yehuda/ Admit** and Dovid exemplified admission. Dovid understood how he was nothing, come out of nothing, going on for infinity and that is why God chose Dovid to be king.

Dovid's entire relationship with God was one of compassion from the very beginning; he was to be a **B'nei Nefila/Miscarriage** until Adom had pity on his soul and gave him seventy years of his own life: Adom lived 930 years instead of 1000 and Dovid lived for 70 years. Dovid understood compassion since his whole being had been dependent upon the compassion of the first man. Dovid understood how God prepared miracles, like Moshe who had been abandoned to the whim of the Nile River, yet was miraculously saved by the pharaoh's daughter.

Dovid's faith was not a matter of dogma (there is no word in Hebrew for the concept of dogma), but rather a clarity in the Torah and a trust in the Creator who had devised him not to exist and then prepared for him miracle after miracle until he was asked by the people to be king upon Yisroel and protect them from their enemies. Dovid knew the greatest protection was in conducting his kingdom according to the Torah. Dovid loved the People of Yisroel and Dovid loved God.

The very existence of the Jewish People was a miracle as was their ability to take the land from the Seven Tribes of a Canaan slaying their 31 kings. With the advent of Dovid, even the Philistines were pushed back into Gaza and the land was at peace. Dovid built a dynasty continuing until today, but he was not allowed to build the Beit **HaMikdesh/Temple** which was dedicated to peace; since his hands had been soiled by blood his son Shlomo from **BatSheva/ Daughter of Seven** would build the Temple. Shlomo whose name means complete also means peace; no blessing is complete without peace.

There was no war in the time of Shlomo and out of peace came verdant land with succulent fruit and abundance as wide as the sky. This was the zenith of Jewish history and from then the Jewish People have been in decline, but what was begun with such pageantry and triumph has culminated in the darkness of a new beginning—Moshiach. The entrance of the Jewish People into their land had stunned the nations around the Mediterranean Sea; three hundred years later in the time of Dovid and Shlomo the land was secured, and the Temple built and the thread to Moshiach had been clearly delineated.

KOHANIM

The son's of Aaron from the tribe of Levi were separated out from the rest of the people and were known as **Kohan/Spiritual Servers**. This term, Kohan, is also used in describing the Jewish People—a people whose purpose is to serve humanity.

Through the Kohanim the Jewish People were able to understand their purpose as a nation among other nations; just as the

Kohanim served God, so too would the Jewish People provide for the spiritual needs of the world. Each nation of the world, like a wife in God's concubine, was the conduit for a particular kind of influence funneled from the heavens to the Earth through the blood of a people; each people make up part of the whole—the Jewish People in conjunction with the Kohanim brought a light to the world from beyond creation, from the **OreAinSof/LightWithoutEnd**.

The soul of Yisroel bound inside the Jewish body received the **Torah/Teachings**. The first word **Anchi/I** engraved the Torah into the heart of the Jewish People. The Teachings were very complex, as complex as the human genome; the human being is a replica of the Four Spirit Worlds divided through the Name YHVH reflecting the three realms of time, past-present-future, plus the three directions on Earth: front-back, left and right, and up and down rendered by the One. The Creator, being completely beyond creation, adopted creation as a language by which to speak to the creation.

The lower three of the Four Spirit Worlds are segmented into triangles depicted in our our solar system as the nine planets numbered from two to ten, like the ancient Tarot cards, then fashioned into the Earth as seven continents and three oceans and lastly engraved into the ground the design of the human body—divided at the hips and between the shoulders and the head. The Jewish People are also divided into three sections: the Twelve Tribes of the people, the thirteenth tribe of Levi who watch and sing in the Temple and the Kohan who serves.

The Twelve Tribes are the legs, Levi the torso and Kohan the head. The head is the most sensitive and the smartest—the head oversees the entire body with meticulous care. The Talmud- Book of Law explains, The purest blood created by the body is dispatch to the head—the brain has little blood but of exemplar quality. One of the jobs in the **Beit HaMikdesh/Temple** was to receive the blood of cows, rams and sheep: the best of the herd—Avraham was compared to a cow, Yitzchak to a ram and Yaakov to sheep.

In order to officiate, the **Kohan Gadol/Head Server** had to wear specific garments as specified in the Torah for the purpose of, *Honor and beauty. And these are the garments which you will make: Breastplate, Apron, Coat, Pants, Settings, Hat and Belt.* The seven garments correspond to the seven parts of the body: head, body, legs, arms and sex. On each shoulder was a stone engraved with the twelve names according to the birth of the twelve sons, progenitors of the Twelve Tribes; on each shoulder was twenty-five letters—fifty depicts eternity.

On the breast plate were gold settings where twelve stones each naturally-miraculously inscripted with the name of a tribe would light up and relay messages from heaven. Through the stones, the Kohan carried the burden of the people upon his shoulders and communicated on their behalf to the spirit world. The cabalists express how the service of the Kohan is relevant to each person, *Without any doubt, in every place we accustom to go we only go to polish and purify the Earth with the letters of Torah and prayer; all of Yisroel are messen-*

gers of the Almighty—each person according the incidences of intersecting realities.

The regular Kohanim wore four garments instead of seven and were auspicious for their meticulous knowledge and agile abilities to juggle cups of blood that should not coagulate prior to being poured. It was the **Kohan Gadol/Head Server** who would sprinkle blood on the curtain separating the interior chamber from the Ten Commandments on the day of **Yom Kippur/Day of Atonement** after entering and offering the incense in this once a year happening. When the Kohan would emerge he would speak the name YHVH and though the people crowded together when the name was spoken there was suddenly room to prostrate on the stone floor.

Greater than all the offerings of animals and produce were the seven candles requiring constant illumination. Like the human being whose life is more than the physicality of being, the point of the Beit HaMikdesh was to project a light into the world; the Temple was constructed with windows wide on the inside but narrow without to project light into the world. The Temple was the lowest citadel in heaven; the next Temple, according to prophesy, will come from the heavens, made of material erodible with pillars of diamonds.

The Kohan was the embodiment of kindness; known for their red hair and blue eyes, the Kohanim were renown for their great beauty and agility. In the Temple, the Kohanim would stand before the musicians from the Tribe of Levi and bless the people; dividing their fingers three fingers left and two fingers right in a way to break the kindness of the light so the people can receive the blessing.

TEMPLE

If the Temple of Shlomo shone with the radiance of the next world, then why was it required to be built specifically from stone? This cabalistic question is at the root of the Temple—in order to radiate into the world the Temple had to be of this world and stone is the most physical, the most inanimate, of all the elements, the paradigm of permanence. The mammoth stones used for the foundation confounds the imagination, particularly given the injunction, No metal object be used on the Temple Mount since Temple's purpose was to promote peace and metal was used in the pursuit of war.

The Book of Kings chronicles the 480 years since the giving of the Torah and the building of the **Beit HaMikdesh/Temple**. Precisely, 480 years after the erection of the **First Beit HaMikdesh/ Temple the Second Beit HaMikdesh/Temp**le was erected in the time of Ezra. The First Temple stood for 410 years while the Second Temple stood for 420 years. The Third Temple which will appear from the heavens will be made of a material which does not degrade and will stand forever with the advent of Moshiach.

The Temple inaugurated the thousand years of **Netzach/Victory** and suddenly the darkness, which had fallen upon the neighboring nations sparkled into life; Homer suddenly arose, writing in perfect Greek the detailed incidence of the war of Troy, which happened five hundred years before. Alliances of peace were being forged as a gentle light spread out from Yisroel. Homer in Hebrew means, The Speaker.

Shlomo, the fifteenth generation from Avraham (the full moon of the Jewish People when her spiritual power was the greatest) was a man of peace and wisdom who brought his people to the glorious zenith with a magnificent building—the first of three steps into Heaven. The many miracles happening in the Temple were only signs; the real secret was in the performance of the actions, as specified within the Written Torah and elucidated through the Oral Tradition—the performance of the 613 Commandments, many of which, could only be done by the Cohan in the Temple.

Of the prescribed events taking place in the Temple, sacrificing the animal was the secret of secrets. The large altar used for the burning of the sacrifices stood just beyond the place where the people gathered, was called, **Ezrut Nashim/Women Chamber**. The **Oulam/Opening** was where the Leviim gathered to played instruments on a platform between the people and the altar. There were numbers set for each instrument except for violins—there could be violins without number.

Further in, beyond the Oulam where the Animal Alter is situated, is the **Hachel/Hall** replete with Golden Candelabra, Table-Bread and Incense Altar leading to the **Kodesh HaKodeshim/Holy of Holies** wherein the Ten Commandments are kept. In the center of the Hachel is a small Golden Altar where a blend of eleven spices prepared in meticulous measures was offered up. The Ark holding the Ten Commandments is likened to the two parts of the brain in constant ethereal connection, stamped with the sign of Shadi by two staves appearing as two Shadiam/Breasts upon the curtain before the light of the candle and the aroma of the bread and the perfume of the eleven spices.

People came from all over the world to offer animals at the Altar and witness the miracles that occurred—through the light of the **Beit HaMikdesh/Temple** illumination, the world came to Yisroel. The Kohanim served God by serving the people—offering up the animals the people brought to be sacrificed; through this interaction between the Kohanim and the people, the people understood how to be Kohanim to the world. All humans have an animal spirit requiring submission; teaching our individual animal to sacrifice for the sake of Heaven is the key to open the heavy doors of Heaven.

God's plan to take the Jewish People down to **Mitzriam, Irva HaEretz/Genitals of the Earth**, to become a people surrounded by the womb of containment—a world without zero, infinity or exactness had succeeded; God took the people out—birthed through the bursting of the embryonic fluid—the parting of the seas, and fifty days later the Jewish People received the Torah with the admonition to go into the land, build the Temple and have a permanent dwelling place for the **Shechina/Presence** of God in the Land of Yisroel.

Chapter Two

Understanding Human History

After the flood and the Tower of Babel the human being became greatly diminished; reduced down to three families (the three sons of Noah) who still retained a measure from the wisdom of the past, soon eradicated by the eschewing of language from the world. The human being was quick to devise an alternative: Cuneiform, a written record of events, tallies of business transactions and the odd story of Gilgamesh imprinted throughout the Mediterranean countries. Scholars say, One of the ways new scribes practiced their art was by inscribing the Epic of Gilgamesh.

Since the flood only covered about fifty feet above the ground, many of the previous monuments like the Sphinx and the pyramids remained as ancient relics pointing back to the past. Recent scientific investigation has uncovered major water erosion at the base of the Sphinx. The flood, recorded in many cultures throughout the world, is a point in time dividing the ancient from the modern–over the last four thousand years, not much has changed. The human being stills struggles understanding the stars and the heavens, unsure of our place amid the vastness.

The past and the future were of great concern to the ancient peoples of the world; if a logical explanation could be found back to the past then a sense of security could be ascertained. The same was true for the future; progressing towards a finite point is reassuring to the human psyche–zero and infinity are the enemies of logic. The idea of nothing is embodied in the concept of something coming out of nothing. This fear of nothing was the crux of the ancient problem 4000 years ago after the great flood; this same problem has perpetually afflicted the world ever since—What is out there?

Nimrod who lived a few hundred years after the flood was able to persuade his fellow beings that the flood was not the Will of God, but an act of nature—not some heavenly retribution. He devised a theory, the heavens moved every 1666 years flooding the Earth, but it was possible to prevent this from reoccurring, cajoled Nimrod—we only had to go into the heavens and put our earthly sword there to protect the Earth. Today, scientists believing in the haphazardness of creation posit the same idea—they are on the look out for meteors headed our way.

So adroit were these ancient people that by combining their efforts, God was forced to retool creation by breaking human language, least they should succeed. Anyone who knew this original language could conduct creation according to their own will and against the Will of the Creator. Freedom of choice had to be curtailed; the one language was broken into seventy languages according to the land and the people—God had prevailed, but there was a price—the

Creator had to tinker with creation. Without a unifying language, miscommunication led to manipulation and violence.

Nimrod began the tradition of manipulation of the human condition and frailties; exploiting the human spirit into relinquishing freedom of choice became the hallmark of tyrannical kings and corrupt governments. Nimrod had built his tower on Mars; Egypt with their ancient relics directed their prayer to Mars; with the help of the Magicians trickery was perpetuated upon the people. Mars was also the War of God worshipped by Rome—Mars recently made his closest approach to the Earth ever in 2003 when Rome was at his zenith. Now Mars and Rome are declining.

Nimrod was known as the first person to hunt people with the power of his mouth, meaning he was the first exploiter of human frailties presenting himself as something his was not while making promises he could not keep. The portal of hypocrisy and politics had been opened on the planet Mars; later in Mitzriam/Egypt, whose name indicates confine (Matzer in Hebrew means confined) whose capitol Cairo is Arabic for Mars, where the beginning of civilization to restrain the human spirit first appeared. The fear of nothing subjected the people to their masters; afraid of the unthinkable, zero and infinity, people shielded themselves by adopting kings who erected monuments to rationality—perfect geometric shapes.

Like a thorn in the foot, the leaders were plagued by the irrational. To make the irrational to be rational, the leaders made a pact with the magicians, who later morphed into religion— together they fooled the people into thinking the world was ruled through the rational, of which the human excelled. There is an inherent fear within the human being of nothing, since there was a time before creation when existence did not exist, relegating the human being into nothing— the essence of the soul fears return to non-existence.

There are Seven Names of God the Torah prohibits from being erased; these seven primordial states bridge the gap between zero and infinity. Each Name corresponds to one of the seven existences prior to creation as recounted in the Midrash-Book of Metaphor. One of these seven primordial existences is the Name Yisroel. The purpose of this primordial Name is to bridge the gap between limited creation and the limitless OreAinSof/LightWithoutEnd. Through service to the world, embodied in the 613 Commandments, the Jewish People were chosen to be Cohanim/Severs to the World.

Existence is made of letters combined into words melded together through a higher supernal energy, giving each separation a unique name. The Name is important because of the intrinsic nature of the name reflects on the bearing of the name. The Cabala teaches how the name is the intermediary between the soul and the body; the human instinct to awaken someone sleeping or passed out by calling out their name is a sign—through the name the soul is conveyed to the body. To God creation is a book, each nature confined the by the letter they occupy.

That which went on forever was as problematic as the dreadful abyss of nothingness. The Name Yisroel, is one of the seven things existing prior to creation along with the Torah/Teaching; **Chesai HaKoved/The Honored Chair** suspended above creation; **Sham**

Moshiach/Name of Moshiach; **Avot/Fathers:** Avraham, Yitzchok and Yaakov; **Beit HaMikdesh/Temple;** and **Tshuva/Return**. According to the Midrash-Book of Metaphor, These seven things help to mitigate the effect of zero and infinity. If the world saw the great value of the Jewish People, each person filled with commandments like a pomegranate is filled with seeds, they would send armies to protect us.

The purpose of the Jewish People is to receive the Torah in each generation and through the Torah teach the world about the nature of creation and how, in the future, Moshiach will bring the world into a new dimension. Through the actions of the Fathers and their children who built the **Beit HaMikdesh/Temple** the peoples of the world can go beyond the rational.

NOTHING

While Egypt was busy building monuments meant to sooth the fears and enslave the populous with a veneer of perfection, in the East zero was taking hold. From the Aryans in India to the Semites of the Middle East, the concept of nothingness was producing astounding results; there is a great advantage in nothing particularly when inferred by the zero.

The Babylonians were first to use the zero in their computations; zero made it easier to write fractions—irrational numbers. Zero embodies the concept of nothing, as demonstrated in the Babylonian system where zero could never stand alone since zero was a number without substance; zero can not effectuate change in any other number by means of addition, subtraction, division or multiplication. Zero is the mother of the irrational from whence is born an infinity of fractions.

The Cabala, received esoteric information, is based on the **Sefer Yetzirah/Book of Form** author by Avraham who explains the foundation for the twenty-two letters of the Hebrew language, yet there is no mention of zero. The alpha-numerical system, which Hebrew utilizes, divides into three sections similar to the human form: 1-9 represent the legs and sex, 20-90 refers to the body, and 100-400 correspond to the head—there is no zero in the human form, but there is the soul. The soul is like nothing, so much so, that many people don't think there is a soul.

There is no thing in zero and nothing happens at zero, but through the zero life is animated—truly, something comes out of nothing. The cabalists are obsessed with explaining this phenomenon, bringing countless examples, describing the difference between cause and effect and the inexplicable spilling out of creation from the void—like clouds suddenly produced from a variance of temperature. The Cabala is physics at the highest level, discerning degradations in spirit. In the **Kodesh HaKadoshim/Holy of Holies** situated in the fulcrum point of the Temple were the Ten Commandments—taking no place, a reminder to the people, Nothing is not to be feared but revered.

The distance between something and nothing, is greater than the distance between one and infinity since infinity is made up of many ones, but how nothing suddenly manifests into something is beyond human comprehension. Ancient people thought, God created creation but then left, leaving the workings of the world in the hands of the planets and stars. The ancients were adroit at manipulating starlight; Avraham came to inform the world, the Creator of the Heaven and the Earth is ever-present for all things all the time. What the Jewish People did was to bring the zero into the one by proclaiming God is One. This strange attribution does not imply first, since the first of many is irrevocably invested and affected by all subsequent numbers and God is not; on the contrary, the Torah insists God is completely beyond any semblance of creation.

ONE

The first letter of the Hebrew alphabet, and also the first letter in the word **Echud/One** is Aleph; Aleph also has the meaning of Military General or Leader. The Aleph is comprised of a dot in the right hand corner, a dot in the left hand corner and a diagonal line passing from the upper left down to the lower right—dividing the two dots. Is one truly two? The pictorial rendering of one into two is perhaps the most important fact of creation; all things can be divided—only nothing is indivisible.

One is an intermediary between nothing and something, as explained in the Talmud, The letter Aleph of the word **Echud/One** indicates the Creator, the **Chet/Eight** is indicative of the seven heavens (the seven worlds from Earth to Pluto divides the space of our solar system into seven segments of heavens) and the Dalet/Four is the four directions upon the Earth—meaning, Echud/One, God is equally available in Heaven as in Earth to all creation. The Temple exemplified the idea, God is One, in the geometry of the building and the service done within; to bring this light to the Earth was the highpoint in Jewish history.

The nature of One combines the virtues of both states, nothing and something by being revealed but the indivisible, much like the corner of an edge or the infinitesimal dot—their existence is totally theoretical. The **Beit HaMikdesh/Temple** built by Shlomo 3,000 years ago displayed these opposite attributes of being and non-being in the apparatus of the Ten Commandments, cubes carved out of stone, but taking no place. Shlomo when ascending to the crown prayed, asking God for wisdom; God granted his wish and as a result he was able to give three hundred metaphors for every idea within the Torah indicating the height to which the king could rise, but no number was greater than One.

The concept of One permeated the Land of Yisroel since all her citizens proclaimed daily the six words of the Shema Yisroel declaring, God is One. The Jewish nation was the light within the void, a light that would bring the world to and beyond infinity. The confine experienced in Mitzriam, when mixed with the Torah, produced a reflection of creation from many levels through the story of the Jewish People—the light coming out of the darkness; confine within confine, in the likeness of an onion, creation was a fruit on the Tree of Life—creation was but one fruit in God's garden, yet God was still One and ever-present.

Everything existed because God existed. God was not of creation and there was nothing within creation that could contain God. Since creation required being brought out of nothing, the nature of creation is to recede back into nothing; without the creation being moment by moment animated by the Creator, creation would easily be devoured back into the void—this intrinsic fear is ubiquitous amid creation. The Jewish People are likened to the wick of a candle immersed in oil, able to hold the flame of fire, bringing a continual light into the darkness of the void.

God is ever-present, God is unseen; God fills the void and surrounds the void with endless light—a small job for God. The illumination brought into the world through the Torah directs life into a future, propelling creation forward—the quest of creation is to go beyond limits, thusly coming closer to God. A balance is struck between the emptiness of the expanse, which is termed *nothing*, and the fullness of the creation suspended between the zero point and the infinity sign; this harlequin creation of Heaven and Earth reflected in the night and day is the stage where God, through the fingers of the stars, conducts creation upon the Earth.

Scientists observe the universe expanding to infinity towards non-existence. On every level of existence, beginning from the atom and going to the limits of the physical universe situated at the end of the four spirit worlds—the void exists, but only as an expanse beyond cause and effect. Beyond time and space is the spiritual world and beyond the spiritual is the void— nothing, incased within the LightWithoutEnd. Understanding the symmetry of creation, the cause and effect, each with a logical inference, helps to clarify our standing in creation. Because God's Will is here on Earth and because we have freedom of choice—this is the center of creation where physical flesh and blood have consciousness.

Zero is a confine as is infinity, but to the Creator they are but one and the same—the confines of creation; everything between zero to infinity in the progression of creation is determined by the free choice of the lowly human being on planet Earth—our influence is small but powerful. In the **Beit HaMikdesh/Temple**, the Kohanim washed their hands and feet from a brass kettle before using the golden vessels of the Temple. Brass and gold look similar but brass tarnishes while the gold remains forever golden—the difference between Heaven and Earth— thus connecting the highest with the lowest, zero to infinity.

The distinction to be on planet Earth in a terrestrial body able to accomplish commandments given by the Creator is important, both on a cosmic level and in our profane daily lives; because God's Will and Intention is on planet Earth, therefore with every thought, speech and action we effectuate the celestial weave of God's artistry—nothing is for naught, existence has meaning woven into the pattern deciphered through ancient texts then illuminated by the Cabala.

THE SECRET

The Cabala explains how an infinitesimal dot (related to the Yud of the YHVH) remained after the evacuation of the light forming the **Mokom Ponoi/Empty Space**. The retraction of the light is likened to a man retracting from woman after leaving the drop of sperm—smallest cell in the body. This dot is the lower dot comprising the **Aleph/1**; within this dot is housed all of creation. God's futuristic Name **Ehiya/Will Be** replaces the **Yud/10** with the entirety of the Aleph/1—as the prophets promised, *On that day God will be One and God's Name will be One.*

This dot of life is represented in the Hebrew letters as the Yud with an accompanying number of ten. Again, likened to the sperm, this single dot of light left from the abundant illumination of the **OreAinSof/LightWithoutEnd** with the Ten Sefirot patterned and fashioned in the form of a human being—the letters of the YHVH when arranged vertically make the stick- figured form of the human being. The dot of the Yud, the first letter of the YHVH, contains the other three letters; each letter houses a spiritual world.

This configuration of three within one is representative of the four spirit worlds: **Atzilut/ Emanations, Briah/Creation, Yetzirah/ Form** and **Asiah/Action**. The World of Atzilut connects directly to the Ore/Light whereas the other three worlds have the **Kalim/Vessels** to contain the light. The Yud/10 is the simplest of forms—a dot, initiating the Hey/5 which looks like a house with two walls, a window and a roof. This primal form of Yud-Hey/15 is the basic component that differentiates man from woman. The male Yud elongates in a Vav/6 a vertical erect line and is received by the **Lower Hey/5** completing the **YHVH/26**.

Placed vertically the YHVH makes the stick-figure picture of a human being. The human being has the capacity to recognize the existence of the YHVH—a Name indicative of the beginning of time when the one dot gave birth to ten; the YHVH is too powerful to create directly and must be filtered through the Name Elohim with an alphanumerical equivalent of 86 the same as HaTeva/The Nature; the Name Shadi with its connection to pi-314 is found by scientists in the most unexpected places. When the values of the YHVH and the Name Elohim are placed together they make 112 the equivalent of Yabok— the famous river dividing dimensions.

Because each letter is made up of infinite other letters called **Milui/Fillings** derived from the spelling of each letter; for example, Aleph/1 is spelled Aleph-Lamed-Pey equaling 111—in turn each of those three letters can be spelled out. Only, two letters contradict the other twenty; One letter has no milui: Mem/40 is spelled MM and is composed of four straight lines in the form of a box; the other letter Samach/60 is in the form of a zero—when engraved into the cubes of the Ten Commandment the Samech is miraculously suspended in space.

These two letters, Mem and Samech, represent the ideas of Infinity—the Mem that goes on and on as if there are no other letters; and Zero—the Samech/60 is the one letter that floats in space and never touches physicality. The Talmud-Book of Law teaches, One in sixty becomes nothing; sleep is considered a sixtieth of death; at sixty a person becomes an elder if empty of ego otherwise he is just an older. Forty, the Mem, is infinity; the fortieth day from conception is needed to grew the placenta—on the forty-first day the child receives the stamp of the YHVH, 41 days according to doctors is when the fetus first takes on human form and the brain begins to develop— until this time the pregnancy is nothing according to Jewish Law, and can be terminated without consequence.

There are two forms of Mem; the first Mem is opened and used throughout but the later Mem is closed and only used at the end of a word—life begins with an open Mem and a closed fist, but ends with a closed Mem and an open hand. The Final Mem/40 like the Samach/60 were suspended within the engravings of the Ten Commandments from side to side of the cube made from sapphire—together forty and sixty equal one hundred.

EVERYTHING

The cabalists insist, inside of everything is *nothing*; *nothing* animates of movement of the atom—*nothing* keeps all things in proper orbit and place. The appearance of something out of nothing is not merely a metaphor to challenge the mind of man and confound the world with the simplicity of the paradox—something out of nothing is a constant reality.

The cabalists explain the concept of nothing by using the brain as an example. The two major lobes of the brain are called Mother and Father; Mother resides in the left lobe called **Bina/Builder**, since from here the emotions form; while the Father is the right side of the brain called **Chochma/Power of What**, since the right side of the brain connects directly from the soul. The soul can not be tested or weighed; the soul takes no place and is not observable at all— the soul is nothing. From the soul, life is conveyed to the body. The right side of the brain sees the nothing and asks, *What?*

What, as in Chochma/Power of What, is the neutral point where the question exceeds the answer. The answer is confined by definition; the Torah never gives definition to the Creator, other than saying the Creator is nothing of creation. Through the nothing we connect to the Creator and through the question in the right side of the brain, we are able to connect to our souls —the nothing binds us. The question opens all doors; whereas, dogma and religion is the antithesis of query. The Talmud-Book of Law quotes, Much have I learned from my teachers, but most have I learned from my students who ask questions.

The experience of *What,* is like being lulled out of sleep by an indistinct something, so beyond anything, it appears to be nothing. When the skies open up and everything which seems real is suddenly obliterated and in the great void and mystery of the cognition standing just beyond the rim of our creation in the greater light the cabalists call, O**re Sovey Col HaOlamim/ Light Surrounding All The Worlds**, then we ask the question, **Mah/What.** This surrounding light, metaphoric to the stars, gives substance to life; as the Midrash-Book of Metaphor says, A blade of grass does not grow until touched by star-light with the message, *grow.*

This surrounding light is the *nothing* pervading everything. The effluence surrounding everything is a powerful light and an uncontainable energy; therefore this light can only surround, yet this surrounding light is the center of everything because everything is based on zero—*nothing*. Since the source can not be grasped by the containment of world; nonetheless this light is still attached to worlds by the virtue of the name, **Ore Sovey Kol HaOlamim/Light Surrounding All Worlds. Malchut/Royalty,** the last of the Ten Sefirot, is the attribute imbued with gravity; King Dovid and his son Shlomo built a magnificent house made out of stone for the Shakina to dwell.

One of the attributes of king, as recounted in the Cabala, is the king has nothing of his own—the wealth of the kingdom belongs to the people and the soul of the king is tied to the Creator for the sake of the people. The relationship between king and people or king and God is one based on love—in love each one tries to give more than what they received. Love enters into the heart and so God gave to creation a second light named, **Ore Mamali Kol HaOlamin/Light Filling All The Worlds**—here we find infinity.

INFINITY

What bothered the ancients about pi and why the structures they built neglected pi is because the irrational was the enemy. People took power over other people by protecting them from the irrational; like Nimrod who convinced people to build a tower to prevent future floods—spiritual insurance. In the Zohar-Book of Secret the sages are often remembered as crying; they would cry before they spoke because the irrationality of what they saw boggled their minds.

The idea of infinity is antithetical to the nature of the human being who lives in a confinement within the limits of time. Infinity is the enemy laughing at the mortality of man; infinity always has more—more time and more space. The beauty of the irrational is going on forever, multiplied by continual divisions until magnificent endless worlds of unique symmetry are formed. We live in a world constantly dividing and separating; dying and degrading. Our world is one of fire with a molten belly drawing all matter back to the essence of heat and ash.

This second light of creation called **Ore Mamali Kol HaOlamin/Light Filling All Worlds** is an illumination able to penetrate creation according to the strength of the vessel. The Talmud-Book of Law states, A full vessel is a weak vessel while an empty vessel is strong. The sacrifices in the Temple were meant to empty the community of the animal inclination towards life; in the empty vessel of the Temple, the spiritual light stream into a stone vessel alive with blood and music and an requite exactness comparable to the human body.

These two lights are likened to the two breasts of the Shechina who dwelt within the **OreAinSof/LightWithoutEnd** from where the Kav/Beam made of two lines of light drawn into the emptiness of creation and then commanded to stop, rendering God the Name, Shadi inclusive of pi. **Shadiam/Breasts**, are portrayed by the two golden-plated staves pressed upon the cloth separating th**e Kosesh HaKodoshim/Holy of Holies**, protruding from the gold-plated box holding the Ten Commandments—the heart of the Jewish People.

Much of life is making ourselves and our surroundings proper vessels to receive the light—empty of physical motives. Light is here in abundance shinning out through the zero dimension of time to bring illumination into each moment. God challenges us continually to open up our hands through our heart and receive the fullness of the light. The truth is, the light is very close; no need to arise into the heavens—the light is graspable through action emanating from the heart.

HISTORICAL PERSPECTIVE

Four thousand years ago, while language was just remerging, pieced together from the shards left from the broken One Language—a result of the Tower of Babel, Avraham appeared on the scene. Time is linear but history is painted on a canvas of finite periods divided into segmentation requiring abrupt changes, until today. There are six segmentations in human history over six thousand years—each thousand year segment producing unique individuals. Avraham came into the world in the beginning of the third millennium; after the 2000 years of **Tohu/Chaos**, God was introducing the two thousand years of **Torah/Teaching**.

Prior to the flood, man was superior in all ways; he was sensitive, ephemeral, possessing great power but out of control—the flood wiped him out saving his gene pool through the family of Noah producing a much more moderate man, in a new era, on the brink of a new segment–the third thousand year period, when Avraham would break through the darkness with the light of God. Avraham was 52 when leaving the Tower of Babel—the year was two thousand; Avraham came to the Earth and through his introduction of holy languages, Arabic, Hebrew and Sanskrit sowed the seeds of holy peoples devoted to God.

The first thousand years of creation, exemplifying the attribute of **Chesed/Kindness**, produced amazing human beings living for a thousand years; then a few hundred years after the flood the second thousand year period, known as **Gevorah/Severity** when the world was destroyed and repopulated, concluded the millennium with language severely broken by the Tower of Babel. The Tower of Babel had been built on Mars, the place from where the flood had erupted. Life migrated back to where it had been in Egypt, cradle of life, where they instituted the god Horus whose strength was insured by the planet Mars with promises of resurrection within an interior world.

Cuneiform, a complex language which could not be spoken, was the only standard of communications used in safeguarding transactions and relating stories like the Epic of Gilgamesh. Hieroglyphics were inscrutable. The remaining oral history had relegated God to a place of secondary importance—God had created life, but left the workings of creation to planets and the stars to rule the circle divided into twelve sections, each divided into six parts equaling seventy-two. There were seventy nations, with seventy languages The two peoples left out of the equation are the Arab People and the Jewish People—two peoples destined to receive prophetic texts, the Torah and the Koran.

God would not only make miracles to proclaim presence in creation, but would take these two peoples and force them into obedience preparing them to receive teachings from Heaven; with this infusion of spirituality the world could become more than the sum total of its pieces.

But first the people must become a vessel for the light. The beginning of Kingship with Dovid King of Yisroel, the Jewish People began their ascent to Moshiach ben Dovid who would then lead the world into the seventh, a final, thousand year period—a Thousand Years of Woman and of Peace. Between Dovid and Moshiach is Jesus and the Koran.

The Jewish People were brought down to the womb of the world to become a nation dedicated to the service of God—to bring light into the world and to the seventy nations of the world. The light is an instruction called, Torah/Teaching, given to the Jewish People on a mountain that smoked and trembled. The experience is engraved into the heart and soul of every Jewish person, the needed inspiration for the conveying the Torah down to the Earth; a strip of land at the end of the great sea was bequeathed the Jewish People as the only place on Earth where all 613 Commandments can be performed—the Promised Land.

The Jewish People returned to the Earth after receiving the Torah; they returned in a cloud that swept across the land bringing them to a narrow pass, the Khyber Pass in eastern Afghanistan heading west to the Promised Land. Historians note an unexplained devastation and crumbling of civilizations along the route recorded in the Written Torah culminating with the Land of Canaan where the thirty-one kings who ruled there were destroyed—replaced by the Jewish People. Historians are baffled as to who these mysterious marauders were but assume the Jewish People were not them.

The Jewish People came to the Promised Land at the same time of the sacking of Troy—a story repeated endlessly, while the story of the Jewish People, as told in the Written Torah, is dismissed. History is simply speechless when it comes to the miraculous; scientifically and stridently, historians are opposed to anything but pure logic—like the ancient ones who feared pi, considering the irrational to be heresy. This ancient divide between those who recognize the Creator as being passive—having created then stepped away leaving the power to the stars; and the other opinion, a minority say, clearly God is active throughout creation. This same divide still remains today.

Not only with the firepower of the Torah did the Jewish People come to the Earth but also with the experience of receiving the Torah—each person felt they had received the entirety of the Torah and insisted on living accordingly. There was no central command prescribing particular actions or ceremonies; everything was spelled out in the Torah, open to individual interpretations as passed down through the families—all the way back to Moshe Our Teacher. An inner cohesion had been formed—whoever was born of a Jewish mother had the inherent obligation to perform the 613 Commandments, the Chosen People.

What kept the Torah intact was the insistence in the Torah that the Torah be lived. The 613 Commandments within the Torah covered every aspect of life—good and bad; whatever problem arose, the Torah was consulted. The Seventy Elders were the Supreme Court and had final say but between the Elders and the people were a myriad of judges with divergent opinions, solving problems and bring about justice. There was little worldly outside influence, since God through the Torah had forbade the practice of other people's cultures, demanding all remembrances to idol worship be eradicated throughout the Holy Land.

The experience of living the Torah caused cohesion and a solidarity beyond language; action was more than words—and though the ultimate purpose of coming to the land was building the Beit **HaMikdesh/Temple**, nonetheless the bedrock foundation are the people becoming a coherent group. So strong was this binding together, a thousand years later when the final remnant of the Twelve Tribes were cast to winds and scattered among the nations, everywhere, the Jewish People remained Jewish; solidified through common action—the 613 Commandments.

The land was conquered divided and settled; judges and prophets were sent and lastly a king was crowned. Now was time to begin the building. Dovid procured the land on Mount Mariah where Avraham had gone to sacrifice his son—a constant remainder as to what degree the Jewish People will go, to serve God. This was the same place where Yitzchok was meditating in the field as the sun set and his son Yaakov lied down in this place, the called a house, and dreamt a dream. Later Mohammed the Prophet would dream, he alighted from this place upon his horse rising into heaven. For God, this was the plac where two brothers kissed.

But, Dovid the king was not allowed to build the building since his life had been hard and his hands were drenched in blood. It would be up to his son Shlomo to build the building, bringing about culmination. The fifteenth generation from Avraham, the full moon of the Jewish People, Shlomo built a building based on his uncanny intrinsic knowing. The crown upon the king's head is emblematic of the crown upon the Torah, crowning the Ten Sefirot with Retzon/Will and **Tianug/Pleasure**; the Temple is the Will of God and the miracles enjoyed there gave the ultimate pleasure—even the metal cherubim were animated.

In a world full of darkness and fear where idols were set up to shield the delicate human being from the ravages of zero and infinity, the Beit HaMikdesh was a light unto the nations bringing the concept of One into a world of abnegation. As the Talmud says, *A full vessel is a weak vessel but an empty vessel is a strong vessel*; the world was full of darkness and the Jewish People were small thereby able to carry the light. Through the **Beit HaMikdesh/Temple**, spiritual light would radiate into the world—thus began the fourth thousand year cycle that would include two Temples and two exiles.

History is not linear and does not proceed from one event to another; history is the record of humanity's progress in the incompressible journey from zero to one, through infinity along a time limited to six thousand years.

Chapter Three

Knowing

THE ANIMAL

Besides being caught between zero and infinity, the human being has to live with the animal; if it were not for the animal, man would give up and take refuge in the zero of depression—the animal within wants to live, at any cost, until the very last moment, the animal hangs on. Without the animal, man would be only ethereal. The animal needs training plus love and care.

In the Torah, animals are compared to angels, having an incomplete knowing, relying on instinctive behavior and living only for the moment. Animals do not reincarnate, perhaps accounting for their strident longing to remain in life. The reason animals do not reincarnate is due to the incomplete nature of their knowing; knowing, more than thinking, represents the human being at our essence. It is to that essence which remains after death. Animals, like angels, are direct representation of the light enlivening their soul; the soul of the animal gets higher in Heaven, closer to the light, according to the use of the animal on the Earth.

The animal soul wants to give, wants to give life onto the Earth for the benefit of the human relationship with God; through the intricate etheric labyrinth used to convey spirit into body, life descends from the heavens into the physical dimension by taking on an animal form. Animals have the unusual quality, given their predilection to life, they are willing to sacrifice themselves for the good of the human being. From the humble mouse servant going around the floor picking up the crumbs to the ferocious lion challenging the courage of man, the animal is here to give, every one a gift from Heaven,

Animals will work themselves to death if told to do so, others will charge into fierce battle to inevitable death without flinching. The animal will be a friend, perpetually withholding judgment. When we die, the animal devours our pieces and though their bodies returns to the earth; the human attributes return to the spiritual essence—the soul. The animal fears death more than pain—the primordial terror of none existence. The harlequin human being made from soul of man and body of animal is unique.

The soul can't quit the animal nor can the animal run away from the soul—they are tied together; the soul longs to be free of creation wanting to melt into the pure effluence of the Creator but the animal, on the other hand, came from nothing and will go back to nothing. The animal has lower expectations wanting only to remain in existence to eat, drink and procreate. God merged animal and human into one body; the struggle between these two disparate forces with opposite intentions is played out on every level of human endeavor.

The diversity of human responses is portrayed in the animal kingdom with endless examples of prototypical human behavior. The Talmud-Book of Law maintains, Many commandments in the Torah would have been learned from the animals if the Torah had not been given. The coalition between animal and man can be depicted as two inmates in the same prison cell who can become companion or enemy. Making peace with ourselves, a prerequisite to world peace, begins by forging a relationship with the animal within. Some have a wild animal, independent and exhilarating, while others are domesticated animals.

Generally, the human being begins from a very young age to meld together for the mutual benefit of both the animal and the soul. Through sacrifice, the animal approaches the soul and through sacrifice the human approaches God. The animals and the angels are dependent on the free choice of the human beings—without this triangular relationship between man, woman and God, creation has no reason to exist.

PLANTS

The Talmud-Book of Law explains, Plants grow from the earth and not from the seed; the seed rots and from the rotted DNA of the seed the ground responds with growth according to the variety the seed planted. The rotting of the seed produces life. This parable of life is extremely pertinent to humanity whose supremacy is in our knowing and our knowledge—wisdom.

Wisdom is a result of rotting; the hard shell of ego breaks, sensitivity is revealed. The flesh exposed to the elements of Earth until only the bones of our essence is left—then comes wisdom. Ultimately, wisdom will purify the Earth as it is written, *The house of Yaakov will go in the light of God; and the glory of God will be revealed.* All flesh will see together—the stones fearing God will speak and beauty will return, as is written, *The effluence and the beautiful crowning the inhabitance of Tavel—Earth.* The Earth is the lowest place in creation, the last stop before returning back is what the cabalist call, **Ore Chazor/Returning Light**.

The Cabala explains, The virtue of return is going back further from where originally projected; return begins in the Earth—growth, life, congnition and self-awareness is part of returning back to the Creator. The Midrash-Book of Metaphor says, *A blade of grass does not grow unless a shaft of starlight touches the blade of grass and says grow*. The animal receives life from the light of the star through eating the grass. When the human being eats the animal, the light of the star, after being filtered through the ground from the plant to the animal, becomes part of human consciousness.

Because the human being is made of the inanimate ground we require these three higher form of life to live; of these three types of life forms, animal and plant rely upon the inanimate earth to make permanent their existence. This symbiotic relationship between man and the elements to procure food and shelter for his family attests to the process of life; each level of life leads into a future with immense rewards—the process of life, living and dying is how we propel ourselves back into the bosom of the Creator.

Everything living moves, but earth moves slowest of all since the earth is closest to the Creator since, The Creator wanted a dwelling place in low—in time and space. Earth is the lowest. All impurity returns to the earth to be cleansed so too does the soul return time and again into the crucible of life to attain purity. The Midrash-Book of Metaphor says, *In the future the rocks of the earth will speak and say—walk on me if you are going to do God's Will otherwise go walk somewhere else.*

The Temple was built of massive stones with magnificent archways and pillars; there was within the Temple a stone altar where upon animals and plants were sacrificed before God, like the engine pushing the process. There are bad animals as there are bad spirits, but none of them were ever seem with the Temple Mount—no scorpions, no snakes and no flies. The Temple is a place of purity where physicality melds with the spirit; fire descended from Heaven in the form of a lion, crouched on the altar then ate up the offerings.

EARTH

The three animated life forms—human, animal and plant corresponding to the basic elements of fire, air and water—die and return to the Earth to be purified. The Earth is unique with continents in the form of a human being of seven limbs: head, body, sex, arms and legs and three oceans likened to the three parts of the brain comprised primarily of water. But, the Earth's uniqueness goes beyond her pleasant veneer. The Earth grows life, and she will grow life forever.

During the Six Days of Creation, as recorded in the Written Torah, all life comes from the Earth; the animals are brought forth from the Earth in the same way plants grow out of the Earth—only the human being was created from the very dust of the Earth. The Midrash-Book of Metaphor explains how dust was gathered from throughout the earth to compose the first human form, so wherever on Earth the person dies the ground will accept the remains. Because the earth grows life, all life returns back to the source—the physical to the Earth and the spiritual into the heavens.

The Earth is phenomenal at giving life and receiving life back—the Earth is always pure, as is the water. Earth and water share the geometry of the circle; the earth is the center of our solar system due to this extraordinary ability to bring about life and consciousness. Though compared to the universe our little planet is but an infinitesimal point hovering by a giant sun which is just a faint spot on the outer rim of a spiraled finger extended from a small galaxy which its self is but an infinitesimal point before the giants of the universe, nonetheless the uniqueness of Earth is without comparison.

Eretz/Earth literally means, ***I run.*** The cabalists explain, movement on this planet indicates movement of the soul. The Earth is a place where we can run; from Earth we get closer to the Creator and along with us everything in the world implored for that purpose comes closer to the Creator. The soul stands in the spiritual realm where the light is strong; the intense light of truth obliterates freedom of choice—coming down into this physical world the soul can move closer to God because the physical can hide the light allowing us freedom to be real.

Fire and air disperse while water and earth absorb and purify. The fact, water draws down instead of merely covering the earth derives directly through instruction from the Creator. The nature of the elements is to align according to weight: earth, water, air and fire, but water draws down as if to go beneath the earth. One of the ramifications of water seeking depth, which is the opposite of her primordial nature to cover, is she freezes upon the surface preserving aquatic life; whereas, earth freezes from within transferring warmth to the surface. The elements are dancing and God is leading the music.

The astounding and beautiful ballet of elements upon the Earth is but a glimmer of the power and beauty of this fourth celestial body from the Sun, comprised of four elements giving rise to four life forms with soul connected to the four spirit worlds built from the Four Letter Name of YHVH. This obvious geometry requires inspection since, through the structure of creation the Creator speaks to creation.

ELEMENTAL GEOMETRY

When the four elements are set against their spiritual counterparts, the four spirit world of **Asiya/ Action, Yetzirah/Form, Briah/ Creation and Atzilut/Close** the number eight arises. Seven is the number of creation as in the Six Days of Creation and the Seventh Day of Rest or as the Zohar contends life exists only on the seven outer worlds, but eight is the sum of the spiritual and physical elements combined—the number of infinity corresponding to the world of **Atzilut/ Close**.

The early cabalists never wrote the word, Atzilut, completely, only the first few letters because **Atzilut/Close** is a world of Oneness; as soon as singularity arose in their minds, the declination of letters was lost. There are letters in the World of Atzilut but they are diminished and absorbed into Oneness, like a candle in the sun. The cabalists say in **Atzilut/Close** the light is one and the vessel is one indicating, each exists separately within the Oneness. The World of **Atzilut/Close,** is where the souls of the Fathers and Mothers of the Jewish People derive. The outer confine of our creation is where existence and Oneness merge in the World of— existence at the perimeter between creation and the infinite array of possibilities from endless **Sefirot/Luminaries** amid the illumination from the **OreAinSof/ LightWithoutEnd**. The other end of our creation is zero. Zero exists only inside the brain of man whose ego pronounces, *I* in the vast emptiness of his imagined zero. The fear of zero permeates the human psyche because the human being is embedded with the complete knowledge of creation—knowing that ultimately creation came out of nothing and could return back to nothing.

Other entities and life forms have an incomplete knowledge of creation and therefore have no fear of zero or infinity; they live in the moment as apparent from there place in the lower three worlds corresponding to animal, vegetable and mineral—to raise into human consciousness, to go beyond the natural fear of nothing, is when God becomes apparent. Nothing and infinity are like the two hands of

God constantly molding creation. The human being lives continually on the brink of both nothing and infinity with only the whim of freedom of choice to keep from the precipice.

In response to the universe, the human being erected their own edifices and monuments defying or appeasing the surrounding forces. Even the people from before the flood with lifespans close to a thousand years had to deal with death's eventuality; the same questions of purpose and consciousness has plagued the entirety of humanity. The box of creation between zero and infinity gave rise to a human being bereft of knowledge as to the scope and position of creation within the plan of the Creator. God did not leave; God is intimately involved in every aspect of life—God creates all atoms in a slightly different place each moment to effectuation motion, bringing about reality.

In the eyes of the ancients, the Creator had fled and left creation in the hands of the planets and stars; wizards saw form in the configuration of the constellations and devised ceremony meant to appease the gods and control the people. If the people were made to believe in zero and infinity, Hell and Heaven, as to the extent of life, they would be easier to manipulate—for this reason, the exploitation of the people, governments and religion made their coalition.

PHILOSOPHICAL GEOMETRY

The Egyptians were the first to use geometry as a way of conveying a philosophical message about life; their pyramids built with exact numbers was a way to refute the idea of infinity, making the box of reality more exact and less mysterious—easier to manipulate. Monuments became philosophies reckoning nothing and infinity; some were tributes to gods, heavenly entities that controlled the forces of nature, while others, attempted to proclaim truth in their structure.

Zero and infinity is a tease to the curious mind, but those who hide from infinity or run to something trying to get away from nothing are obstructions and deterrents to the evolution of our planet. To understand the meaning of life, there is a natural philosophy embedded within the configuration of creation describing how something came into being out of nothing in two different perspectives, from zero to infinity: God is something when compared to creation which is nothing in comparison to God; so God sees creation as nothing came out of something—God. The human being sees the opposite—something, creation, came out of nothing—God.

The cabalists explain how *nothing* is actually sandwiched between God's perspective and man's perspective—each sees nothing. God sees the nothing of creation which came out from God when God created, while man sees the nothing produced the something of creation. Nothing is a buffer between God and creation—what creation views as infinity God sees as nothing; the nothing beyond infinity God sees as endless contractions diluting, constraining and distilling the light until eventually able to come down to the physical Earth. Likened to the sun radiating endless rays of light, one of which strikes the earth, God's radiance is similarly dispersed throughout creation. The Midrash-Book of Metaphor states, To God creation and its continual sustenance are but a small accomplishment.

The zero separating God and man is commonly called, spiritual. The physical accoutrements accompanying man are for his survival: the animal, the plants and the goodly Earth surrounded by a myriad of stars inhabiting endless spiritual worlds. The segmentation of the four spiritual worlds took place because the Creator had taken on a name of four letters, the YHVH, therefore everything divided into four until the physical plane where life became manifest on the forth orb out from the Sun the center of the solar system. The symmetry of creation is meaningful—repositories full of ancient wisdom.

Along with four was ten; each of the four worlds was comprised of ten thus the number forty became predominant. The number forty is expressed as the letter Mem, which looks like a box and is spelled out MM having infinity instead of an endless **Milue/Filling.** The flood rained upon the Earth for forty days and forty nights; Moshe went up to the mountain for forty days and forty nights; the Jewish People wandered in the wilderness for forty years. The placenta grows for 40 days before becoming a child through the stamp of the soul in the form of the YHVH.

Fifteen generations after Avraham came to the Earth, fleeing the Tower of Babel, Shlomo son of Dovid erects a Temple whose configurations mimicked the configuration of creation. Fifteen generations is indicative to the fifteenth day of the month when the moon is full and a great light shines down upon the Earth.

TEMPLE

God's response to the world of idols and edifices was to send the Jewish People to the Earth with the Torah. Besides revealing great secrets of creation, the Torah also commands the Jewish People to erect a house according to specific measurements and other exact specifications. The Temple was to incorporate the workings of the Mishkon where during the time of Moshe a fire would descend from the heavens in the form of a lion and eat up the sacrificed offerings. This was repeated in the Temple.

The basic geometry of the Temple was a rectangle opening up from concentric squares beginning with the cube of the Ten Commandments placed into a wooden box covered and lined with two additional boxes of gold; the entire apparatus was housed in a square room called **Kodesh HaKodeshim/Holy of Holies**—the head. A rectangle extending eastward was divided into three chambers: **Hachel/Palace of Meeting**, the more interior part where the Golden Candelabra, Golden Table and Golden Altar for the burning of spices were serviced by the Head Kohan; **Olom/Hall** where the sacrifices were carried out by the other Kohanim; **Ezrat/Room** where the people gathered.

The philosophy implicit in the Temple celebrates the human form in as the intermediary, spiritual space between God and man—zero and infinity was replaced by geometry. Pi was celebrated in the personage of Shadi, the Name God used to create creation. The Temple was where one went to experience God beyond creation and through the experience melded man to God. Many of the 613

Commandments can only be performed within the Temple; the Kohan performs the services for the entire people—through the Kohan all the people take part.

The Temple differed from the Mishkon in one important way—its construction was in opposites. The Mishkon was assembled on foundations of golden sockets to house the vertical wooden beams with a roof of animal skins. The obvious progression from inanimate, the golden sockets, to plant, the boards of cedar, to animal, the red skins of the bull and the multi-colored skin of the Tachus/Unicorn, as described in the Talmud-Book of Law. Emphasis was put on the mobile animal since the Mishkon was meant to be dismantled and moved.

The Temple was the opposite, made out of stone with the permanence of the Earth, this was the pinnacle point of the Jewish People—to make a permanent dwelling place for God on Earth. But, more than the permanence of the structure or the splendor of the garments worn by the Kohanim were the commandments performed in the majestic halls built from massive stones untouched by metal blade. King Shlomo introduced the builders of the Temple to the *Shamir* a worm with the power to cut the mighty stone and so the Inner Temple was built without the sound of iron—symbol of war; and none of the larger stones could be cut in earshot of the Temple Mount.

Once erected, the Temple became the heart of the Jewish People, who brought thanksgiving offerings for joyous occasions or guilt offerings because of transgressions; the people brought the grain from the fields in the spring and the fruit from the trees after their first three years of growth and gave part to the Temple for a sacrifice. To everything there was a prescription from the law; the blood offering meticulously conveyed through the golden vessel from the animal to the altar, so too was there a continuous stream of activity going in and out from the Temple—only the king was allowed to sit in the Temple.

The altar where the animals were slain and burnt was a hub of activity; the Kohanim, who served in the Temple, interfaced with the people. As prescribed in the Torah, the people were commanded to bring sacrifice in the form of an animal: sheep, cow, goat, turtledove or flour offering and it was the job of the Kohan to officiate.

SACRIFICE

Besides being an edifice to the human form, the Temple was also indicative of the two Names: YHVH and Elohim representing the two degrees of *nothing*. The first revelation of nothing happened when God revealed illumination in the form of YHVH replete with Ten **Sefirot/ Luminaries**, but to God the *nothing* was the creation. The YHVH once exposed in form became the first revelation of nothing from the Creator. The restraint and contraction required to produce a form out of no form was considerable, but the illumination of the Name YHVH was still too potent for creation's own good—creation would be unable to stand the light.

The second degree of nothing was in the form of the Name Elohim with a gematria of 86 corresponding to the word HaTava/

The Nature. Viewed as a word YHVH has the sense of bringing into being; the four letters of the YHVH can be rearranged to form the three tenses of time all in one word: **Hoya/Past, Hova/Present and and Yehiya/Future**. The Name Elohim is plural providing a pathway for the One Light from the YHVH to be manifest in a plurality of ways; as if YHVH is the Sun and Elohim is the moon manifesting the sun's light in a variance of ways.

Nature is the buttress between Heaven and Earth disguising miracle with rationality and instinctual behavior. The nothing behind nature, human being suspects; the fear of living on a flat plane and falling off the edge—the end of anything is the end of everything. Nature was secure and predictable, but unsettling was the fact, nature came out of nothing and goes on to infinity.

The cabalists explain Elohim as being the shield placed before the sun of YHVH so creation could exist; no one can look at the Sun, but the moonlight is the light of the Sun attenuated upon the Moon before being delivered to the Earth.

The Temple is a place merging the Names YHVH to Elohim and through the offering of natural sacrifices, brings humanity closer to the essence of the light. YHVH was exemplified by the Ten Commandments kept within a golden box that was measurable yet took no place within the dimensions of the room it occupied; the Ten Commandments were shielded from the rest of the Temple by a curtain protruding like two breasts from the outward press of the staves jutting out from the sides of the **Aron/Box** which held the

Ten Commandments—the hint of pi with its promise of infinity stamped in God's Name **Shadi/Enough** with the gematria 314, pi.

Outside the **Kodesh HaKodeshim/Holy of Holies**, housing the Ten Commandments, were the elements of physicality. To the right was light, the seven candles of the Golden Candelabra-sight, to the left were the twelve loaves of bread on the Golden Table-sound, and in front was the Golden Altar-smell, where incense was offered; the three represent aspects of thought manifested from the two lobes of the brain, right and left, plus the power of knowing related to the cerebellum in the back center of the brain opposite the forehead.

Animals were slaughtered and brought to the large alter filled with dirt and covered in metal where the meat, fat, blood and organs from the sacrifices would be offered. In between these two areas was the sound of the song and instruments of the Leviim—the tribe from where the Kohanim derive, the sons of Aaron, brother of Moshe who received the Torah. The Talmud- Book of Law recounts how from the time of Moshe and until the time of Shlomo fire descended from the heavens, crouched like a lion and devoured what was offered up. The meaning of this phenomenon is found in Yechezchiel's famous prophesy.

Three animals—lion, ox and eagle—lift the chair which carries the head of man. These three animals, the chair and the head of man hover above creation; the three animals are representative to wild animals, the lion; domesticated animals, the ox; bird and fish are sourced in the eagle—the animals lift the head of man higher. The head of man is where complete knowledge resides, connecting to the

Creator through the spiritual realms, but the only way to elevate and approach the Creator is through the animal, the vegetation and the metal of ground.

The human being with his freedom of choice who sacrifices the earthly produce which grows and feeds the animals and the wild beasts. The Jewish People were commanded to sacrifice their life-force to God. What one experienced at the Temple was the metaphoric manifestation of God and man merging in ceremony and miracle.

MIRACLES

The Talmud-Book of Law cites ten miracles occurred in the Temple of Dovid and Shlomo: 1) A woman never miscarried due to the smell of the animal flesh; 2) The flesh never rotted 3) A fly never appeared where the animals were slaughtered; 4) The Kohan Gadol would never have a wet dream the night before **Yom HaKippurim/ Day of Atonement**; 5) The rain never put out the fire of the altar; 6) The wind never altered the pillar of smoke from the sacrifices; 7) The bread offered was never spoiled; 8) The people who stood pressed together would have room when they required to bow and prostrate; 9) A snake or scorpion never attacked within Yerushaliam; 10) No one ever said to his friend, *I had a bad time where I stayed in Yerushaliam.*

Yet, it was not the miracles that inculcated the people with awe and obedience; rather it was the genetic experience of 15 generations culminating with the construction of the Temple. Miracles can convince those without belief, because one believes in what one does not know; but faith is beyond belief, since faith is in what is already known. The miracles were present to remind the Jewish People to strive to go beyond nature; miracle transgresses the laws of nature opening the world to endless possibilities.

The miracle is God's way of saying, *come closer*. The sacrifices in the Temple were called Karbonot built from the word, Karev/ Close. Particularly, through the sacrifice of the animal, the human being comes close to the Creator; because the animal knows no bounds, within instinctual boundaries, given enough freedom, the animal will break down walls with relentless effort—like grass eats up cement. The animal's scant ability to know, allows them to exceed the human; when these two forces are melded together the caldron of life has accomplished purpose. Like the animals of the prophesy holding the chair and lifting it higher, so too terrestrial animals lift the human experience.

Miracles are not proof of God's existence; the Jewish People need no proof—God's presence is implicit in life. Miracles challenge the human spirit to go beyond nature, beyond the dividing emptiness; to plunge into the bosom of God. Dovid records in the book of **Tehilim/ Psalms,** composed and compiled into five books saying, *In my haste I said that all men were liars.* Dovid said this line because, Shmuel had anointed him king but Shual had chased after him and was moments away from slaying him when a miracle occurred.

Written in the Torah is, *If a man comes with miracles but teaches to go against the Torah—kill him.* The Torah, the miracles and the prophecies were all secondary to the action. The Torah is an explanation of the 613 Commandments to the Jewish People for which they were chosen; the miracles are a sign to go even beyond one's nature in the performance of the 613 **Mitzvot/Commandments**; the prophecies served as rebukes to straighten the path of the people. Miracles do not prove anything; miracles are the opposite of magic—magic is done through the hands of man while miracles come from God, through Heaven to Earth.

During the five hundred years when the prophets existed, dating from the time of Shmuel and Dovid, literature disappeared from the Middle East. The light from the prophets obliterated the musings of man and there were more prophets then noted in the 24 books of the Torah, but there were hundreds of thousands of prophets throughout the Land of Israel and beyond. oracles appeared throughout the world as spiritual light funneled down to the Earth. The time of Shlomo, who built the Temple, was a time of world peace—the greatest miracle of all. The Talmud-Book of Law states, ***Shalom/Peace*** *also means completeness since without peace no blessing is complete.*

PROPHETS

The Torah is based on prophecy and Moshe was the first prophet. The Rambam, compiler of the law 1000 years ago, explains the scope of prophecy and the laws governing what is prophetic. The prophet had to prove through three positive prophecies; negative prophecies were meaningless since repentance could always overcome prophecy, but a positive prophecy was a promise from God, which must be kept in exactness.

The Rambam gives an example, if the prophet prophecies the end of a draught saying, *God has promised rain tomorrow from sun-up until sundown* and it rains all day but ceases an hour before sunset—kill the prophet, for he is false. In general, the Jewish nation carried out capital punishment very rarely, invoking the saying, *If the courts put to death more than one person in seventy years the courts were called murderers,* but for false prophets the courts would go out of their way to put them to death since the entire Torah was built upon the veracity of prophecy.

There were schools for the prophets since the time of Shmuel where those with a proclivity for the spirit would go and study with the elders; the young prophets would make themselves vessels through merriment brought on through music. Bands of prophets wandered throughout the land rejoicing in song and dance, at times throwing off their clothes to prostrate naked on the ground reveling in the spirit. The prophets are referred to as **Meshagoim/Crazy** by the working folks in the Book of Shmuel, because of their antics. Nonetheless, the prophets were an essential part of life, able to locate lost items and see into the essence.

Mothers would bring their children to the prophets to ascertain the nature of the child. The Talmud-Book of Law relates the story of a woman who was told her child had the nature of a thief. She asked what could be done and they suggested the child always wear a hat to remind him, God is always watching. For this reason the custom for

Jewish men to cover their heads, least they should forget about God, is prevalent. The Talmud insists, Most thieves are Jewish, therefore mandating the custom of covering the head. Women are not obligated to cover their heads due to their natural connection with God.

These five hundred years of prophecy were not confined to the Land of Yisroel or the Jewish People; oracles were set up in Greece and Egypt drewing people from around the world, but the prophets of the Jewish People were different since what was written down directly related back to the Five Books of Moshe. Not all prophecy was written down, only that which was required for subsequent generations, most prophetic happenings were only relevant during a short time or period. For five hundred years the eyes of the prophets were open, then the Temple was destroyed and the king was blinded.

Four books are added to the Torah from the times and lives of the prophets. Yechezchiel with his etheric visions of sublime creation, Yeshia with his Messianic message rampant with signs of culmination at the judging of the nations; Yirmiyaho whose name implies bitterness and who rebutted the people, but to no avail and was force to watch the carnage of the Jewish People and the destruction of the 410 year old Temple. Twelve other prophets are included in the portion of the Written Torah known as the **Naviim/Prophets**; the last third is called **Katovim/Writings**, some of them written by the kings.

Though the prophets spoke the word of God exact, the people did not always listen because there was a king and if the king did not listen the people did not listen.

KINGS

The prophets acted as spiritual advisers to the kings, but the kings had no obligation to listen to the prophet. The prophet brought spiritual advice or admonition concerning the people and also brought independent messages for the king.

Nason HaNavi had been prophet for Dovid, Shlomo and Cheskiyaho, a righteous king who foresaw in his future a wicked child who would lead Yisroel astray and so decided not to have children; when he became ill the prophet Nason HaNavi arrived announcing the king was going to die because he did not do the will of God.

If so, said the king, *I will have children.*

Too late, replied the prophet. *The heavenly decree has been signed.*

Cheskiyaho turned to the wall and prayed to God to heal him and he would go forth and have children; the king's health returned—the spiritual decree had been annulled by the King of Creation. Cheskiyaho sired Menasha who set up idols and turned the people from following in the way of Dovid and Shlomo. The full moon of the Jewish People was fading, the darkness of ego and jealousy was slowly creeping across the face of the light; the Jewish People were destined to decline and become but a sliver of light yet out of the darkness of exile would come Moshiach—the new moon.

Throughout the 450 years from the time of Dovid, the turmoil of kingship and domination riddled the Land of Yisroel; good kings like Asa, Cheskiyaho and Zdchayaho destroyed the shrines to Baal, but wicked kings came after them erected new idols. Prophets were sent, sometimes at the risk of their lives, like Eliyaho who went so far as to break the spiritual law offering sacrifice outside of the Temple, in order to bring the Jewish People back to the way of Torah. The prophet went out into the world and challenged the magicians and sorcerers who were erecting idols. Eliyaho won.

Without doubt, since the days of Dovid and Shlomo, the high-point of the Jewish nation, things began to decline, taking over 400 years. The two books of **Meleckim/Kings** and the two books of **Dvrie HaYomin/History** recount the events, following the blood lines and recording its details of how the Jewish nation lived the Torah. History is not incremental but conforms to the ether of time, each thousand year period, rising than falling to another; the fourth millennium, the known as **Netzack/Victory** would eventually fall into **Hod/Retreat** with the destruction of the Second Temple two thousand years ago.

In the end, the Jewish People, as decreed from heaven, having been warned by the prophets, were sent off into exiled with the Temple destroyed and the people disgraced, the Land of Yisroel was left fallow for seventy years making up for the 70 Shemitot/seventh year over the last 490 years when the people of Yisroel did not let the land rest. The Talmud-Book of Law who relates this fact also teaches, The destruction of the Temple and the exile of the people happened because, though the Jewish People lived the Torah they did not love the Torah.

The Jewish People were being held to a very high standard as the Midrash-Book of Metaphor says, God creates and destroys constantly until arriving at the desired end. By destroying the edifice of

Oneness which bridged the fear between zero and infinity and taking the people from their land decimating their numbers while inculcating new cultural attitudes produced a Jewish People who loved the Torah but hated each other. There is much more to the Jewish story, the story of exile and the story of redemption, again and again throughout a 3500 year-old history.

Jeremiah writing in the book **Aicha/Lamentations** compares the Jewish People to a divorced women—her house torn down, forced to leave the land of her husband and wander aimlessly in the world. And, as Dovid prophesied in the book of **Tehilim/Praises**, *On the shores of Babylon there we sat and cried remembering Zion/Stone.*

MOSHIACH

At the end of Shlomo's rule, as written in the Book of Kings, the old king's wisdom turned into folly; his excesses became too excessive—he had too many wives, too many horses and he returned to Egypt and took for his wife, daughter of the Pharaoh. As a result, God made a decree and sent his prophet Achiya Hashiloni to inform Yeruboam from the house of Yoseph of the heavenly decision. The two men left the city and went into a field because Yeruboam and the prophet were the statutes in Torah like a man compared to the grass in the field. Yeruboam was the greatest of his time.

The prophet brought a mantle of royal cloth, with which he covered Yeruboam, then removing the garment, the prophet proceeded to cut the thing into twelve distinct pieces saying, *Take ten.* The tribes of Yehuda and Benyamin were thus separated from the other ten. Dovid and Shlomo were from the tribe of Yehuda; the Temple was built on the adjacent tribe of Benyamin on the border of Yehuda, God was ripping the kingdom into two, one centered in Yerushaliam in Yehuda and the other in Shomron in the portion of Yoseph, causing Moshiach to be also cut asunder into—Moshiach ben Dovid and Moshiach ben Yoseph.

In the end, for all his ascension and refinement, Yeruboam turned bad, erecting idols and encouraging the people to serve the idols; he set up two golden calves where people could burn incense—he even went so far having sex with his mother. She asked him, *Do you have pleasure from this?* He answered, *No, I do this to insult God.* Through the apparatus of Heaven, God challenged Yeruboam, *If you will return to me and to the ways of the Jewish People, then I will let you walk together with Dovid and I.* Yeruboam asked, *who will go second?* God replied, *Dovid*—Yeruboam refused to repent, if that meant going after Dovid.

The Rabbis say, Yeruboam had too much arrogance; the Cabala explains, the attribute of **Malchut/Royalty** requires humility—even if the person warrants being king, his arrogance can destroy everything. As a result, the wicked king was doomed to return to Earth and reincarnate as Moshiach ben Yoseph—a zealot fighting for God; later Moshiach ben Dovid would come and bring peace. The story of the Jewish People is the story of Moshiach, how world peace is achieved by a higher understanding of the Torah; Moshiach means, Oil Pourer—oil raises above water. Often the Torah is compared to water, coming from the heavens but retaining purity upon the Earth—Moshiach will bring oil.

Oil happens as a result of breaking, squeezing, compacting until oil is emitted—oil naturally elevates above water. Oil is also easily combustible; fire is the opposite of water— water draws down but fire draws up. Moshiach will bring oil into water and upon the face of the water will be the sparkle of illumination. The existence of Moshiach in the world began with Yehuda and Tamar centuries earlier when God decided the time had come; 500 years later Moshiach became refined into the lineage of Dovid, now the spark of Moshiach exists around the world.

Division made the kingdom weak and broken; politics became more important than people, the squeeze was on—divisions multiplied as history forged a decline in the light from the moon, the pressure was building. When Moshiach comes, along with a new understanding of the Torah, will be a new Temple descending from Heaven made from indestructible material with pillars of diamonds. The First Temple was the seed, the Second Temple was the rotted seed and the Third Temple along with Moshiach will be the Tree of Life.

Chapter Four

Kindness

LOVE

The word **Ahev/Love** in Hebrew is based on the word **Hev/Giving**; the added Aleph gives the sense of, *I give.* In Jewish Law, the act of marriage requires a man to give something of value to the woman; if she accepts the gift, then she is married—the act of giving and receiving binds the two lovers. Incumbent upon the man to extend love, so the woman can receive; after receiving the love from the man she is able to then respond since man's love comes from a more external source—whereas, a woman's love is internal.

The Torah is the gift from the Creator to the Jewish People; the receiving of the Torah on Mount Sinai was the marriage ceremony obligating the Jewish people and all their descendants to honor only the God of Avraham, Yitzchok and Yaakov. Land was purchased for the bride and a beautiful house was built so God and the Jewish People could unify through the 248 limbs and 365 veins mirroring the human form related to the 613 **Mitzvot/Commandments** made of 248 acts of obligation as acts of commission and 365 restrictions required. The performance of these decrees simulates love between the Jewish People and the Creator of Heaven and Earth.

Some commandments are fulfilled in thought, like one's intention or the turning of the mind from the prohibited desires; others commandments are performed through speech like, the recital of the Shema or giving truthful witness—but, most of the commandments are accomplished through physical action. The same holds true for love-making, one's intentions and words are implemented by the not so subtle acts of the body. The 613 **Mitzvot/Commandments** are representative of a proxy body for God who has no form. The Jewish connection through the Torah as a conduit with the Creator, like a rope between Heaven and Earth, allows a new light entrance into the world.

As expressed in the logic of thought propelled by an ethereal Will sustained by endless ecstatic Pleasure from the **OreAinSof/LightWithoutEnd** the **613 Mitzvot/Commandments** is how the Jewish People touch God. The history of the Jewish People is framed in the tale of a celestial love affair between Earth and Heaven, where the Jewish People and their love for God is the thrust of the story. Incumbent upon the man, according to Jewish Law, to sexually satisfy the woman, so the Creator built Yisroel a beautiful house on a pristine land lush with succulent fruits, delicate wine and thick golden oil. It was a land flowing with milk and honey.

The years of the marriage between the Jewish People and God is reflected in the connection between the lower seven Sefirot divided six and one like the outer seven planets where life exists according to the Zohar-Book of Secrets. From the standpoint of Earth revolving around the Sun in 365 days the marriage lasted close to a thousand years, but from the perspective of the Creator seen in the 248 year cycle of Pluto, the ninth planet from the Sun, the marriage lasted only four years and then the husband announced to his wife, You don't love me anymore and therefore a divorce is in the future.

Shlomo King of Yisroel had written a book called **Shir HaShirim/ Song of Songs** detailing the relationship and love between God and the Jewish People through the metaphor of a strapping young man of princely benevolence tasting the lips of his beautiful bride and praising them as being greater than the finest wine. Four hundred years later the prophet Yerimiya would write, using the metaphor of a mature women being divorced by her husband, the book of **Aichaa/ How?** Detailing the destruction of the Temple and the exile of the Jewish People .tantamount to divorce.

How is it that I sit alone? A vast people like a massive widow amid the nations, taxed into servitude. I cry and cry, in the night my tears are upon my cheek and there is not comfort from all who love me; all my friends torment me and are to me as enemies.

DIVORCE

According to Jewish Law divorce is enacted through a written statement signed, dated and witnessed releasing the woman from any obligations to the man. God left it to Nebachanezer, King of Babylon to deliver the papers, destroy the house and send the Jewish People out of the land. The Temple had stood, not as a monument or an edifice, but as a place of worship for 410 years where the commandments were carried out scrupulously, but it was not enough—the people lived and learned the Torah, but they did not love the Torah.

A few centuries earlier the Ten Tribes had been vanquished by Assyria; at first Nebachanezer, King of Babylon thought the Temple walls were too thick and fortified to breach but after awhile he noticed a change; he asked a soldier to go to the wall and investigate—sure enough each day the wall became smaller. The King of Babylon understood all he had to do was wait. When the time was right he took the city and destroyed the Temple. The day was the ninth day of Av—the month of the Lion when the First Temple fell.

God left the Jewish People to wander in the world and the Shechina went with them; the chance of returning back to the land was slim but in the end after seventy years a miracle happened and two tribes returned, Levi and Yehuda. Ten tribes were lost amid the turmoil of the nations. Levi returned to rebuild the Temple, so the sacrifices could be instituted, but Yehuda's return was more than a tribe it was an attitude of deep-seated faith.

The word Yehuda in Hebrew means: to admit. Admittance is usually used in relationship to guilt and certainly the Jewish People had what to be ashamed of having numerous wicked kings. The Jewish People were being chastised because they did not love the Torah; those who admitted to their disgrace of having held back love from the Creator were considered to be **Yehudim/Admitters**, no matter the tribe. Yehuda also means, Acknowledge the Oneness of God, even in the darkness of exile. The Jewish People were likened to a widow, because God had turned away and now the love of the Jewish People would be tested. Exiled from their land and their ceremony, they still held fast to the study of the Torah and performance of the commandments.

The 24th book of the Written Torah, Devri HaYomin/ Chronicles ends thusly, *God of their fathers sent to them in the hand of angels daily admonition because of the love of the people and their difficulties. But, the people laughed at the angels, despised their words and disgraced the words of the prophets until there went up God's anger upon the people until there was no recovery. There went up upon them the king of Chasdim who killed the young men in the Temple and had no compassion on the young or the old and the withered all were given into his hand.*

All the vessels of the Temple, big and small, the storehouse of God and of the king and his servants were all taken to Babel. The Temple was burnt down and the walls around the city were torn down, towers were burnt with fire and all the precious things were destroyed. At the point of the sword those who remained were taken off to Babel. And it was to them and their children to be subjects of Persia to fulfill the word of God in the mouth of Yirmiyaho: until the land is appeased from the **Shmita/Seventh Year of Rest**—seventy years.

In the first year of the reign of **Charesh/Cyrus** king of Persia there went out a voice in all the kingdoms and was written saying: So says Charesh/Cyrus king of Persia, all the kingdoms of the world are given to me YHVH God of Heaven has appointed me to build the

Temple in Yerushaliam which is in Yehuda—who is like them in all the peoples? YHVH, God is with them and they should return and ascend to their land.

NEBAKANEZER

Nebuchadnezzar, King of Babylon, the faithful prince appointed by the will of Marduk, the highest of princely princes, beloved of Nabu, of prudent counsel, who has learned to embrace wisdom, who fathomed their divine being and reveres their majesty, the untiring governor, Marzahn, Joachim. Inscription on The Ishtar Gate,

The king's allegiance to Marduk, an ancient name invoked by many of his predecessors, is an allusion to Nimrod—the root of both words, Marduk and Nimrod, is **Mared/Rebel**. Nimrod was the first to rebel against God convincing peoples of the world, The flood was merely a natural occurrence happening every 1666 to be preventable by building a tower into the heavens. By establishing a kingship in the heavens to protect the Earth, Nimrod's real intention, to war against God, could be disguised.

Fitting, Avraham's enemy, Nimrod, who tried to throw him into the fire, would be the appointed messenger to burn the Temple and exile the Jewish People through the reincarnation into Nebachanezer. Trying to invoke his predecessor's success Nebachanezer also built a tower in a valley and a massive idol whose end, like that of the Tower of Babel was to fall to the ground. Nebachanezer chose Daniel, a child prophet, to write the wisdom in the language of Chasdim, the ancient language dating back to the Tower of Babel.

Daniel was chosen by the king because he was a Jewish child with very fine hair; the hair hinted at the child's exceptional mental acuity. The root of the world Chasdim is Shadi with the gematria of 314—the beginning numbers of pi. Another King of Babel wanted to make war with God. The Book of Daniel relates, The king, Nebachanezer was cursed by God to eat grass for seven years; historians note the absence of writings from the king of Babylon for seven years. The king eventually threw Daniel to and his three companions to the lions then into a cauldron of fire; in both instances a miracle intervened—Nebachanezer began to learn about God.

Nebachanezer is referred to in the Torah as Ariel the same nomenclature given to the mysterious animal who fell from Heaven in sparks of fire during the time of the First Temple, crouched at the altar and ate from the offering. The Midrash-Book of Metaphor states, one lion devoured another lion—Ariel is the essence of the Lion. The lion is related to the planet Mars known as the previous world of **Tohu/Chaos**, the place of unbridled Will and Pleasure. On the E+arth, the animal is bridled and trained for a higher purpose than animal instincts.

Nebachanezer destroyed the Temple on the Ninth day of Av, the month of the astrological lion as is said, *When the lion roars everyone is afraid*. The cabalists explain, The roar of the lion is the revelation of **Keter/Crown—Pleasure and Will**. The Jewish People were not able to receive the power of Keter until Nebachanezer came and took them into exile. The bitter exile in Babylon would push the Jewish People to go beyond their confines of the Law and enter into the metaphor through which would bring the secret, that drop of oil, from where Moshiach will come.

DANIEL

Perhaps the most stunning words in the Book of Daniel are at the end because the words and visions relate to our history and the present time dilemma.

*In the first year of the kingship of Darius of Persia that I stood to help him; and now I tell the truth to you—there will be three kingdoms of Persia (*Cyrus, Darius and Xerxes*) but the fourth will be very wealthy and strong and be called* ***Yavon/Greece****. And there will stand a great king and the multitude will do according to his will (*Alexander the Great*), but his empire will break into the four directions and no other amid his ranks will rule in his stead.*

The southern kingship (Middle East) will be strengthened and they will rule multitudes. At the end of the appointed years will gather the daughter of the king of the south (Esther) will go to the northern king (Xerxes)and make peace with his ministers; but she will not retain power nor will her seed stand—the one who brought her and fathered her (Mordicai) will strengthen her in the time.

There will stand up a sprout from the root that will come to the soldiers of the fortress of the Northern King and be strength to him (Mordicai and Xerxes). *And also Elohim was with the anointed ones (the Jewish People) and the vessels of the Temple of silver and gold should be returned to Yisroel to stand against the King of the North* (Darius of

Persia). *The kingship of the Southern King* (Jewish People) *return to their land.*

The children of the north gathered for war with many soldiers and swept across the land from their strength. Bitter was the kingship of the south and they went out to battle with the kingship of the north who would given over into his hand but he lifted his heart and myriads fell and the south (Yisroel) *was helpless.*

The Northern Kingdom (Rome) *returned with the original multitude and ended the time* (of Yisroel) *with a great army. In those times of oppression the distain of your people will try to stand up the vision* (Moshiach) *but will stumble. The Kingdom of the North will push their battering rams against the fortified city and the arms of the south will not be able to restrain them; the Chosen* (Jewish) *People will not have power to stand.*

They (Rome) *did to them as they wished and they* (the Jewish People) *could not stand before those standing in the land of the deer* (Yisroel) *destroyed in their hand. They set their face with arrogance throughout their kingdom and with their princes they did bad; and the wicked ones of the Jewish People gave them what they needed to destroy so the Jewish People could not stand up and would not exist.*

He (Rome) *turned to conquer the Isles of the Sea* (Great Britain) *and conquered many, but the kingdom* (of Rome) *was torn down from their leaders reproached by their reproaches. They turned within to the strength of the land stumbled, fell and were found no more. In place of power stood the tax collector* (Paul) *and in a short time the kingdom was broken—not in anger or in war.*

In the place of power stood the despicable person (the kingship of **Arom/Naked**) *who did not possess the benevolence of kingship and cast the world into darkness. The arms* (of the church) *will flood, all that is before him and all will be broken—accompanied him will be the wicked of the Jewish People.*

From the influence of the wicked of the Jewish People he will learn and practice trickery; their power will strengthen and multiply because of the little nation of Yisroel. With the promise of peace he will come into the fat of the land and will do what as none before him—pillage and rape and destroy spreading upon the confines of the earth until their time is expired (2000 years).

His heart will awaken with strength and will challenge the Southern King (Islam) *and will engage in battle* (Crusades) *of vast armies strong beyond comprehension but they* (Islam) *will not stand because of the divisiveness* (of Rome). *They eat bread with the nation with whom they are set to destroy from their flood of soldiers the multitude will be slain.*

And the two kings (Rome and Islam) *their hearts deceiving upon one table, they speak lies and what they devise will be of no success since their appointed time expires. They will return to their land* (after 2000 years) *with great wealth and their hearts determined to deal with the Jewish People returning them to their land.*

At the appointed time they will invade the Southern Kingdom (Islam) *not like the first time or like the last time.* (The First Persian Gulf War lasted from 1980 to 1988—Iraq against Iran. The Second Persian Gulf War began in 1990 ended in 1991) *Coming with mighty ships and fortifications they hit and stayed to occupy to do what is wicked to the Jewish People; they stayed and built resentment on the Holy Covenant, building their empire of those who abandoned the Torah—Holy covenant.*

They will disgrace the Name and the knowledge setting abomination in its place. The wicked of the Jewish People will be corrupt with flattery while the people who know God will be strong in what they do. The intelligent will inform the masses, but they will fall to the sword and the flame in captivity disgrace for years.

In Jewish stumbling the Jewish People will be helped a little, but many will join with the flatters (religion). *Those of intelligence will stumble to align with them, to explain and excuse without end. The king* (of Rome) *will do as he wishes elevating himself beyond good and powerful God; he will speak foolishness yet he will prosper until indignation tears him down that which he built.*

On the God of our fathers Rome does not understand the delicacies of women—he does not understand the concept of God or all that is great. To God he is a disgrace that he should stand in a place of honor and pray to a god that his fathers did not know, he gives the offering of gold, silver and precious stones; he will do for the strong who acknowledge the new strange god giving much honor to the rulers of the Earth who will divide the spoils.

In the end will rise up the people of the Southern Empire (Islam) in storms that overcome the King of the North; with horses and chariots and many ships they came and passed over the land coming to the land of Yisroel and many stumbled—these are the rescued from the hand of Edom (Rome) *and Moab* (Persia) *and the first children of Amon*

(Egypt). And a hand was extended throughout the world to receive the (Jewish) *refugee.*

He will have dominion over the Arab economy, Africa and others will fall into step. He will hear rumbling in the North (Russia) *and the East* (China) *and he will be afraid; he will become angry to destroy and exterminate. He will put his tents upon them between the sea and the mountain and will come to an end and none will come forth to help him.*

In this time will arise an angel of kindness who will stand with the Jewish People during their difficulties, which no nation has suffered like the Jewish People, each one a written book; a multitude from the Earth will awake and will live forever and some will awake and be forever ashamed. Intelligence will shine like a light from Heaven to help the masses recover and be righteous like the everlasting stars.

KETER/CROWN

The Midrash-Book of Metaphor states, When the Jewish People received the Torah on the mountain Sini, the angels brought two crowns and crowned each person with the Crown of Doing and the Crown of Understanding. When the Jewish People no longer merited the Temple the two crowns were taken from the heads of the Jewish People and given to the world: one crown went to Babylon and the other to Rome.

The cabalists explain the phenomenon of **Keter/Crown** as, The convergence of *nothing. Nothing* is found where two opposites meld, for example a corner; each wall is complete in itself, but the nothing of the corner allows them to merge at an angle. Similarly, the *noth-*

ing of **Keter/ Crown** merges Heaven with Earth through the duality of **Atik Yamin/Ancient Days** and **Erech Onpin/Large Faces,** as described in the Zohar-Book of Secrets. These strange nomenclatures were later identified by the cabalists as Pleasure and Will—encompassing energy unrelated to a specific organ or limb.

Pleasure and Will are the determinative factors in regard to freedom of choice since true freedom can only exist beyond the confines of intellect. Pleasure and Will both exceed the boundaries of logic; considered chaotic, their source is in the previous world of Tohu/Chaos where unbridled desire was expressed through narcissistic ambitions. Logic sorts out erratic wants to better satisfy the inner longing. Pleasure and Will are neither good nor bad but rather the essence of all emotions necessitating logic as an intermediary. Rulers, who are naturally attached to the heavens being responsible for masses of people, are relied on for their Will to do what is right for the people.

There is no historical explanation or even salient theory as to why Nebachanezer stormed into Yisroel, destroyed the Temple and carried the people away into exile, but from the standpoint of the Jewish People this was the ebb and flow of the ocean against the sand. When the Jewish People would return seventy years later, a people greatly diminished from twelve tribes down to the two but first they had to be thrown into the crucible of *nothing*—having nothing, being nothing. Bitter is the exile "Crying on the banks of Babylon".

The obligation to do the 613 Commandments and the accompanied laws extrapolated from the study of the Torah had not been removed from the heads of the Jewish People, only the crown of glory, which was nothing. The Jewish People brought the essence of the **OreAinSof/ LightWithoutEnd** out to be revealed in the world, to the nations. The Jewish People now knew God in the sleep of the exile; a nightmarish dream seeming to go on forever—one kingdom rose and another declined in the constant churning of nations until the crown came to rest on the head of *nothing*, incarnate in the world—Haaman, from the nation of Amalek.

During the time of Dovid, the Prophet, Shmuel slewed the king of Amalek after he had spent a night in captivity; during that night the king had managed to impregnate a slave girl whose seed was watched and bred for centuries until giving birth to Haaman. Now, the Persian Empire ruled, having defeated Babylon, and the king Xerxes ruled in the city of Shoshan. In Shoshan is where Mordicai and his wife Esther lived. They had left in the exile which brought them to Shoshan. Mordicai was an elder and Esther was young and beautiful.

AMALEK

The nation of Amalek is considered in the Torah as: **Rashist Amim/ Head of Nations** referring to the Seven Nation of Canaan who lived in the land promised to Jewish People by God to Avraham—father of three nations. Canaan had been cursed before his birth by his own grandfather, Noah, to be a slave to his brothers; so when he left Africa, the portion allotted to Chom his father and went into the portion of Shem his uncle, God said, Let the slave hold the land until my people arrive.

The Seven Tribes of Canaan who populated the land of Yisroel represented the seven bad emotions derived from lower seven sefirot of the Ten Sefirot, which broke in the World of Tohu/ Chaos and seeded this world with life; the seven day week corresponds to these seven emotions—in increments of seven the human being through invoking freedom of choice, can dramatically change the world—turning bad into good. The intention of the Jewish People leaving Egypt was to enter into the land and turn the bad into good—but Amalek attacked the Jewish People and tried to kill them while yet in the infancy of their nation while leaving Egypt.

The Midrash-Book of Metaphor compares this attack on the Jewish People to a man who jumps into boiling water; though he gets scalded his action opens up the ability for others to follow after him. The cabalists confer on the nation of Amalek the distinction of being the will and intention animating evil; Amalek attacked the Jewish nation right after her birth, cutting off the weak ones—the sign of Amalek is to attack the children and sodomize the weak. Claiming the human being is perverse in sexuality they would cut off the genitals of the dead and cast them into the heavens.

The cabalists contend Amalek to be a none-entity, an evil essence attacking human will and intention, separating the heart and the head with an iron gate—the gematria of Amalek is the same as **Barzel/Iron**. One of the 613 Commandments is to, Remember to forget Amalek. God further announces, the Name of God will be divided as long as Amalek exists. The Talmud-Book of Law dictates, whenever you find Amalek—kill him. Both, remembering to forget and killing Amalek, requires motivation, action and aggressive behavior, because Amalek is aggressive.

Amalek forces the will to submit through the arrogance of pride and boasting; the nature of non-existence is to arrogantly exert presence, subduing the natural inclination of the soul towards good actions—the power of Amalek is to inflict doubt. When a person doubts the presence of God, the will is left to meander. Amalek, head of the Seven Nations (the seven bad emotions) casts doubt into what one knows and through the doubt ignites a strange fire within the heart of man. Doubting the presence of God, brings about a plethora of bad actions.

The seven bad emotions are the same as the seven good emotions; emotions have no morality but are a delineated display of the attenuated Will, having been filtered through the intellect of the brain. Amalek is the manifestation of pure Will without purpose other than to exist; like a bubble in a stream, temporal and bloated, so too the perspective of Amalek is a magnification of nothing. But the force of *nothing* is formidable.

HAAMAN

Haaman came from the nation of Amalek embodying the arrogance and pride natural to his species who knew amongst other things how to shape-shift, often taking on the visage of a lamb—thus the saying, A wolf in sheep's clothing. Amalek exceeds hypocrisy—a type of subtle shape-shifting; the hypocrite through discovery and embarrassment can be made right, but the non-entity of Amalek can only be forgotten and thus destroyed. When you see Amalek within yourself, kill him.

Haaman's rise to power was sudden and somehow related to the episode of the late Queen Vashti who had refused her husband's wishes to appear naked before the king's court. and thus threw the empire into sudden chaos. If the queen did not respect her husband the king, then wives would have the presumption to disrespect their husbands. This dilemma was not allowed to fester, the king executed the queen and sent a decree throughout his empire which extended from Ethiopia to India—127 nations under his rule, women were not to disrespect their husbands. Vashti was the granddaughter of Nebachanezer, a way for the king of Persia to make alliance with Babylon.

Perhaps, the dispatching of the queen made a spiritual breach through which Haaman ascended to be second to the king, but this also marked the end of the Babylonian Exile. After the matter with Vashti was finished, the king got lonely, so his advisers suggested the king gather the most beautiful women throughout his kingdom and sleep with one each night until he found the one he liked and make her queen. In was in this atmosphere, Haaman became elevated to his post second only to the king decreeing, everyone had to bow down to Haaman.

The Talmud–Book of Law explains, The Torah would be received in completeness a thousand years after the original giving on

Har Sinai/Mountain of Hatred because there was too much revelation at the inception. The Book of Esther, where is recorded this final receiving, is distinct from the other books of the Torah—God's Name is never mentioned. The name Esther relates to the word **Hester/ Hidden** because the miracles occurring were hidden in coincidence. The practical reason why God's Name was not invoked is because the Jewish People were still in exile under the Persian king and the story had to be composed to aggrandize the ruler.

The wicket Haaman received his power by relying upon chance; he was famous for his lottery by which he would affix dates to carry out events at prodigious times leading undoubtedly to success. He had a devoted wife who had born for him ten sons and many friends who loved him; he was widely admired and it came as no surprise Haaman was elevated above the princes to take his place at the side of the king where the pomp of royalty suited his arrogance. The Torah says, *God brings them high so God can bring them low.*

As God had used the Seven Nations of Canaan to hold the land, and had exploited Pharaoh at the birth of the Jewish Nation by hardening his heart, bringing about the Ten Plagues; so too, did God use Haaman from the nation of Amalek to test the Jewish People and their love for the Torah and devotion to God. When God gave the Torah, the people heard the lightning and saw the thunder; during the time of Esther all anyone saw was politics and the struggle for power—even in this environment, the Jewish People clung to the Torah and to God.

The Midrash-Book of Metaphor describes the giving of the Torah as God holding the mountain over the Jewish People saying, Accept the Torah or I will drop the mountain. The cabalists explain the mountain in the metaphor was representative of the revelation of illumination during the giving of the Torah prohibiting anything but obedience. Later when the light was hidden, would be the true test of the Jewish People. Purim, the holiday associated with these events, means Lottery in the Persian language.

ACHAEMENID EMPIRE

A short lineage of Persian kings is the historical connection between the Babylonians who destroyed the Temple and the Greeks who would be next to conquered Yisroel and the world. The fulcrum of the Achaemenid Empire was between Cyrus the Great who defeated Babylon and released the Jewish People from slavery and Darius who returned the Jewish People back to their land, Yisroel, to build the Temple.

In between Cyrus and Darius was Xerxes King of Persia who lived in seclusion allowing the cruel Magi to rule in his stead, the wicked Haaman. Though the Achaemenid kings were enlighten, ... *I am Cyrus, King of the world. When I entered Babylon... I did not allow anyone to terrorize the land... I kept in view the needs of people and all its sanctuaries to promote their well-being... I put an end to their misfortune. The Great God has delivered all the lands into my hand; the lands that I have made to dwell in a peaceful habitation...*

The reclusive Xerxes preferred the pomp and ceremony while letting the Magi rule, but the king feared losing power; so when the seventy years had passed since the destruction of the Temple, the king gave a lavish party—since, obviously the prophecy of the Jewish People returning to their land after 70 years had proven fallacious. Xerxes brought out the golden vessels used by the Kohanim in the Temple to serve the needs of the people who came into the garden of the king where the party persisted for six months.

When at last the people had been served the king made another party for the princes and at the end of the party commanded his queen Vashti to come forth and show her self naked, but she refused. The culmination of a year long party left the king with a dilemma requiring a special edict to the people insuring men remain the head of house—causing the king to introvert even further. Vashti had been a marriage of alliance; he no longer needed her—she was expendable. However, a king without a queen is not really a king, according to the Talmud-Book of Law.

Loneliness brought the king forth in search of another queen; his harems were aplenty with beautiful woman and each night he would have a new and different experience but nothing satisfied the longing of the king. The eunuchs who served in the harem suggested a decree be put forth throughout the kingdom to encourage young women to come to the king and perhaps be made queen; the zealots went out and took women—Esther was taken to king from her husband Mordecai. In the Book of Esther, for the sake of peace with the ruler, the story reads, Mordecai brought his niece to the king.

For a year the young women would be kept under lock and key while preparing for the king. For six months the woman would bath and perfume her body and for six months she would be adorn with beautiful things until finally the night would arrive to be with the king. Whoever the king did not choose to replace the former queen would be relegated to the harem for the rest of her life. This continued until Esther went into the king—the king had found a queen; she was beautiful and full of grace and the whole kingdom rejoiced.

The Book of Esther recounts these deeds; read on the holiday of Purim considered the greatest of Jewish holidays, the scroll of Esther is a foretelling of the future—our time when the revelation of God is muted and hidden. According to Herodotus, the famous historian who recounted a similar story from the mouths of the Persians, writes, *The Persian people consider this day the greatest holiday and celebrate it solemnly. They call the holiday, Beating of Magians.*

MORDECAI

The royal party happened seventy years after the Temple had been destroyed by the king of Babylon and the killing of Vashti represented the beginning of the return. Mordicai had been witness to the destruction, he also carried the memory and knowledge of the Jewish People and when the king of Persia announced that he would give a party in celebration of the destruction, Mordecai understood the Jewish nation had arrived at a defining moment—did the Jewish People doubt prophesy? Haaman from Amalek was the purveyor of doubt.

Mordecai tried to dissuade the Jewish People from attending the party but the sumptuous orgy amid ornate opulence was hard to resist. The Jewish People were in jeopardy of melding into the Persians and losing their Jewish identity. True, Cyrus the Great, had allowed 40,000 to return to the land of Yisroel, but under the rulership of Persia, the Temple was not to be reconstructed, thereby thwarting to the desire for the Jewish People to return to their land. What the people needed was a miracle—the miracle of Purim.

The party's subtle acknowledgment that the prophecy remained unfulfilled was inspiration for the Jewish People to meld into his empire and relent on their dreams of returning back to their land and rebuilding the Temple to reinstitute the sacrifices. Mordecai understood the exile to be like a cauldron, a place to purify the Jewish spirit—many would be lost to riches and freedom of Persian life, but those who remained in their Jewish identity would merit return, the building of the Second Temple and the reinstitution of the 613 Commandments.

At the climax of the party, Queen Vashti who also gave a year long party for the women, teaching them all sorts of ways of the body, was called forth, but she began menstruating and declined. Mordecai saw in the incident the first glimmer of light, the long exile was ending seventy years from the time of the destruction of the Temple in Yerushaliam.

Mordecai knew the time to act had come, but what should he do? When the decree went out from the king for women to be brought to the palace so the king could requite another queen, and the guards came and took away his wife because of her beauty, he saw his chance through his beloved Esther—he commanded her, to go with the other women but not to tell of her heritage. Esther did what she was asked trusting in Mordecai's wisdom and after a year of preparation was brought before the king to spend a night.

During her year of preparation Esther found favor in eyes and hearts of those who met her and, no surprise, when the king decided to marry her and crown her queen. Though the event certainly was a harbinger of happier times, the road back to the Land of Yisroel would be fraught with great dangers—that became evident with the ascension of Haaman bred from the seed of Amalek. The heavens watched to see who would be rendered victorious and who would fall? Would the Jewish People really receive the Torah or would they falter?

Everyday Mordicai went to the palace gates just to be close to Esther, hoping to hear some word justifying, having his wife taken from him, his beautiful Esther into the harem of the king, but now become queen. Because Mordicai was a Jewish Elder he was obliged to know the seventy languages derived from the breaking of the Tower of Babel; through his knowledge of language, he overheard two guards from the inner part of the palace planning to harm the king. Sending a secret note, through an intermediate, he informed Esther of the plan.

The king investigated, found the plot to be true and was written in the chronicles of the king—Mordecai, from the people of Yehuda, had given true witness. The two traitors were hung and no more was said. But, Mordecai was not favored by all, since he refused to bow before Haaman, who made a practice of wearing an idol around his neck—Mordecai refused to bow before an idol. Finally, someone pointed out to Haaman, Mordecai the Jew refuses to bow down; Haaman's fury could not be satisfied by killing one human being—he looked into the heritage of Mordecai and decided to kill him and all his people.

ESTHER

Haaman was the head Magi, whose power stem from the magical powers inherited from Amalek; he knew it would not be easy to destroy the Jewish People, since even the great Bilaam in the time of Moshe had been unable to curse the Jewish People—eventually Bilaam was killed by the sword. Haaman would have to be smart; by relying on the power beyond logic, surmised Haaman, he could conquer a people whose foundation was logic.

He would throw a lottery of dates; he arrived at the 13th day in the month of Adar—the last Jewish month of the year, after winter, at the precipice of spring, almost a year hence. A year would give ample time to notify the 127 country-states under Persian rule from Ethiopia until India to exact the punishment of extinction on the Jewish People on an appointed date in early spring. Runners were sent to all provinces stamped with the signet ring of the king, which Haaman possessed—having ransomed a fortune of gold into the coffers of the king.

When Mordecai came to the palace gate dressed in sackcloth with ashes smeared on his forehead, a sign of mourning, Esther was alerted. Through a series clandestine messages, Esther took upon herself to approach the king and beg for mercy for her people; since she had done as Mordecai had commanded and not revealed her ancestry, she had the best chance of dissuading the king through the element of surprise. She and her ladies in waiting would fast for three days before approaching the king; Esther also asked the Jewish People to fast for three days—2500 years later, the custom is to fast the day before the holiday of Purim.

The miracle of Purim is hidden in the successive events occurring in quick order. First, Esther approaches the king without permission; whoever approached the king without permission would be killed unless the king reached out his golden scepter, which is what the king did, asking her, What she wanted so badly as to put her life in jeopardy? Esther replied, she wanted the king and Haaman to come to a little party she would to give later in the day. The king is quick to command Haaman to come to the queen's party.

They come but Esther refuses to reveal her secret until the next day when she begs her two guests return. Haaman is feeling particularly exhilarated, having been so honored; his feelings soared until he leaves the palace and sees Mordecai wrapped in sackcloth, unwilling to make even a gesture at the tyrant. When Haaman returns home, sharing his good news with his wife, his ten sons and many guest who came to celebrate with him, he has to admit, his happiness is blunted by having to see Mordecai wrapped in sackcloth.

Why wait for the appointed time? Asked his wife, *Why not go to the king and ask permission to kill Modecai, now!* And that is what Haaman did—going early in the morning, hoping to get an audience with the king. The Talmud-Book of Law states, The miracle of Purim happened that night when the king could not sleep; the cabalists explain, This is metaphoric to the King of the Kings who could not let the Jewish People suffer and die; God could no longer pretend to sleep in the darkness of the exile.

The king asked to be read the chronicles of his kingship. When they read how Mordecai had informed the authorities of two errant guards who were plotting to kill the king, the king asked if anything was done for Mordicai for his service? Nothing had been done. The king then asked, *Who is in the courtyard.* Haaman had come early in hope of having an audience with the king to ask permission to hang Mordecai the Jew; Haaman was escorted into see the king. Haaman had prepared a gallows **50 Amot/Measurements** high on which to hang Mordecai.

Haaman, from a line of wizards, knew the importance of seven; he understood, Seven times seven, forty-nine, was inclusive of everything within the realm of logic, but, fifty was beyond logic, where good and bad were equal. Because of Haaman's evil intention, the rabbis later instituted the requirement in the celebration of Purim to become intoxicated to the point of being unable to distinguish between good and bad as a way of rectifying the bad intentions of the wicked—**Shtut D'Kadusha/Stupid for God.** The Torah was given on the fiftieth day after leaving **Mitzriam/Egypt**, after the seven emotions times seven had been purified—ultimately, the Torah is also beyond logic.

When the king asks Haaman, *What should be done to one who the king wishes to honor?* Haaman, thinking the king meant him, suggested the king's crown, cloak and horse be given to the man whom the king wishes to honor and a dignitary to take him around the province announcing to the people, *This is what the king does to those whom he wishes to honor.* Haaman was surprised when the king asked him to do those things for Mordecai the Yehudi. Haaman had no recourse but to obey the king.

When Haaman returned home, with a bent head and fallen face; after telling what had happened, his wife announced, *Once you start to fall against the Jews, you fall.* As she spoke, the guards from the palace came to escort Haaman to the party Esther had prepared with the king— he was full of trepidation. Esther was going to reveal her secret request to the king, Mordecai had just been hailed a hero, his family saw a bad omen coming forward and Haaman could do nothing but wait and see what would transpire.

After some prodding, Esther explained to the king, She was from a people who had been taken from their land to wander in the Earth and if only to be sold into slavery, she would accept the king's decree but to be slaughtered mercilessly by the sword was more than she could bear. The king wondered how this was possible and asked Esther, *What person did this*? She responded, *The evil Haaman.* The king became upset, he left the room and the party and went into the garden. When the king returned, Haaman was leaning on Esther's bed pleading for his life.

The king looked and saw; rage threw from his mouth—*first you take my kingdom and now you try to entice the queen?* One of the eunuchs who accompanied the king whispered, Haaman had recently erected a gallows 50 Amot/Measurements high. That is how Haaman and his ten sons were hung and the anger of the king assuaged. The book of Esther is the only book in the Written Torah that does not contain the Name, or even any reference, to God. The light Moshiach will bring will render the holidays a thing of the past, except for the Book of Esther and the holiday of Purim—a time relevant, forever.

Chapter Five

Severity

METAPHOR

The Talmud-Book of Law states, the seventy years spent in the exile of Babylon constituted the repair needed for the Land of Yisroel since the Jewish People had neglected their obligation to let the land rest every seven years. The cabalists explain the number seventy is related to the seven lower Sefirot each comprised of ten; Haaman and his ten sons comprised the physical manifestation of Amalek depicting the **Keter/Crown** of Will and Pleasure

Amalek is a non-entity, a will without substance waiting to fill a vacuum with pride and arrogance. Like a virus, Amalek is able to disguise, appearing to be like any other organism, then unleashing its stealth of malignant disease, trying to kill the host. The virus, like other sicknesses, is used by God to exact punishment, bringing about renewal and redemption; sickness is an instrument ready to be unleashed—a concept explained by the cabalists as T'shuva/ Return. When sickness occurs, the natural response is, *Why me and why now?* Asking these questions is the beginning of **T'shuva/ Return-Repentance.**

In the Torah there are two types of darkness: a darkness of substance like the night sky or a darkness, which is merely a vacuum bereft of light; in the vacuum, Amalek is powerful, sucking real life into the vacuum giving substance to non-entity existence. Once within the clutches of Amalek, doubt seems reasonable; sickness and difficulty evoke dark emotions articulated by a malicious inner voice telling the person, God is not here; God does not care— **T'shuva/ Return** is the process of reuniting to the light by casting off the darkness of doubt.

The first step in **T'shuva/Return** is to leave where you are and return back to where you were; for the Jewish People, the land of Yisroel is our eternal destination. The demise of Haaman allowed a respite and as a result of the queen's intervention and the revelation of her ancestry, Mordecai was elevated to a high political position and Esther's son Darius was in position to acquire the crown—which is what he did. According to Jewish Law, a Jewish Soul can only be transmitted through a Jewish woman—Darius was Jewish and would be the catalyst for the return of the Jewish People to the Promised Land.

When Haaman wanted to kill the **Yehudim/Admitters** he was not referring to one of the Twelve Tribes but the entirety of Jewish People by destroying their faith in God; those who admitted to their unique relationship with the Creator of Heavens and Earth, understanding their Jewish Soul obligated them to perform 613

Commandments, were selected out for elimination while the others were allowed to assimilate. In this way, ten of the Twelve Tribes were lost; these are the same ten tribes who separated from Yehuda and Binyamin shortly after the erection of the First Temple.

Only those who admitted to their obligation to God and to the land, longing to rebuild God's Temple, returned. Exile had accomplished an ongoing clarification within the Jewish People; the stalwarts and the zealots—all by creative design. Reminiscent to the goats of atonement—one goat was to be sacrificed in the Temple (those who returned) and the other goat (the ten lost tribes) was to be sent out into a rugged terrain to die in the wilderness. The division occurring to the Jewish People accomplished a dissemination of ideas throughout the world along with the attenuation of the light.

This attenuation of light continues until Moshiach because as the cabalists insist, *The greatest light comes out of the deepest darkness.* The ebb and flow of exile would have a long and difficult terrain, impossible to follow or understand if not for the metaphor.

THE STORY

The Midrash-Book of Metaphor states, When the Torah was given to the Jewish People, God lifted the mountain above them and said, *Accept the Torah or I will drop the mountain.* The Rabbis understood, the receiving of the Torah had been done under exceptional circumstances and the true reception would not be completed until later.

The cabalists explain, The giving of the Torah on Har Sinai was accompanied by the light of revelation, rendering impossible, resistance to the Torah—as if a mountain were held over the heads of the Jewish People threatening to drop should they have any inkling not to accept the Torah. Later, in the time of Esther, after a thousand years of regal history in the Land of Yisroel, now lost in the exile of Babylon, the real receiving of the Torah could be accomplished. God's Name does not appear in the story of Esther because the receiving of the Torah was done regardless of spiritual revelation.

Thousands came to the call of Mordecai, who had been instructed by Queen Esther that the Jewish People should fast for three days before she would meet with the king. These were the **Yehudim/ Admitters** who were ready to give up their lives to be Jewish; many others just took their lineage into the world and melded into the bloodline of other peoples—the **Yehudim/ Admitters** received the Torah. God promised at the giving of the Torah, If the Jewish People accept the Torah than I promise to bring Moshiach by the year 6000.

For those returning to the Land of Yisroel, the travail lasting 2600 years had begun culminating in the world giving birth to Moshiach and the subsequent 1000 Years of Woman and Peace. The story of how the Jewish People barely escaped extinction at the hands of Haaman would be replicated countless times throughout the world—miracle without revelation. Through the miraculous story of the Jewish People, God's Light enters into the world for all to see—a precarious position for the Jewish People, but God wants to be known throughout the world and the Jewish People are the vehicle for God achieving God's intention for creation.

In the story of Esther, which is the story of exile, God's influence is inferred in the most subtle way, the king is unable to sleep—from insomnia, the miracle of Purim stems. Purim provides the story of our perpetual exile until our mission is accomplished—the bringing of Moshiach through the spreading of the light. Shia Islam has historic roots in Jewish teachings— particularly regarding matters of Moshiach. The Jewish influence in Persia went to the very soul of the Persian People because, the king himself was Jewish. Darius allowed the Jewish People to return and rebuild the Temple.

All in all, five exiles are counted: Egypt, Babylon, Persia, Greece and Rome; the culmination of the Roman exile, which has continued for over 2000 years, will result in the coming of Moshiach; the story of how Moshiach happens is yet undetermined. Each exile provided a barrier between what was before and what would come after—each time the light of Torah was attenuated again; accompanying the exile back to the Land of Yisroel was the acknowledgement, The time of prophecy was concluding; the codification of the written word— the 24 books of the Torah, would ensue.

The Written Word was only part of the story; in between the lines was space for endless detail mapping of mathematical concepts wrapped in esoteric garb. Within the confines of the Written Torah was everything that could be learned, but the trick was in the extraction. Some keys to the Torah would come from abroad in the form of **Geirim/Converts** and slaves inculcating themselves into the Jewish People; from the **Geirim/Converts** came great names like Rabbi Mier and Rabbi Akiva, Onkelous and countless others; the cruel Herdot, the freed slave— according to Jewish Law, A freed slave automatically becomes Jewish.

The codification of the 24 books of the Written Torah means, The book was completed but the story continues.

WRITTEN WORD

Ezra and Nechemia were the last two prophets who led the Jewish People back to the Land of Yisroel; Ezra and Nechemia constitute the end of the 1500 year-old prophetic story beginning with Avraham, Yitzchok and Yaakov—the three Fathers. The Twelve Tribes, whose gestation and birth into the Jewish nation came from the womb of the Earth, **Mitzraim/Confine (womb of the Earth),** took 430 years; after receiving the Torah and forty years wandering in the wilderness, followed by 400 years settling the land, before culminating with the building of the Temple at the end of the third millennium—**Tiferet/Truth**.

The 410 years of the Temple was followed by seventy years of exile as prophesied by the Prophet Jeremiah, along with the subsequent return of a remnant of the Jewish People back to the land. The last book of the 24 written books of the Torah, **Divri HaYamin/History**, is a thorough genealogy from Adom and Chava until, *So says Choresh king of Persia; all the kingdoms of the world have been given to me by the God of the heavens who appointed me to build for God the Temple in Yerushaliam which is in Yehuda—who is likened to them from all peoples? God, the people ascend.*

The darkness of exile was not over, only mitigated by circumstance and a refined Persian culture who respected all peoples, but on the horizon was Greece. Reestablishing the Jewish People back onto her land was not easy; the Book of Ezra says, *During the day we built, but at night with spears we watched.* No longer did the light of spirituality permeate the Land of Yisroel, instead the dichotomy between day and night, light and darkness, was seeping into the pure water of Jewish ways; after seventy years the Jewish People were considered strangers on their own land—idolatry and witchcraft was prevalent.

The new Temple, a modest structure compared to the magnificent building Shlomo had erected was lacking essential elements, the room housing the Ten Commandments was empty—no cube, no golden case, no cherubim and no staves protruding like breast against the outer curtain and the miracles were no longer present. The Temple had become an edifice where sacrifices were carried out and procedures followed, but the soul of the Temple was not present, only the stone. As if the Temple was asleep in the gentle folds of the hills, the blood of sacrifice spilled upon the stone, the light of consciousness weakly lit while the smell of bread and incense waft into a delicious aroma—but, the essence of the Temple was missing.

The pleasant recollections of the past beyond the painful memories of the recent exile lead to a thirst for the 24 books of the Torah. The number 24 in Hebrew is written as **Cof-20/ Daled-4** together make the word **Cud/Cup**. The collected recollections of the people, the oral portion of the Torah, filled the cup to overflowing; the wise men, who dedicated themselves to the study of the Torah, were able to delineate the truth through constant debate—whatever did not contradict the Torah, was Torah—part of the teaching.

In the Torah, contradiction is an outlaw and the many inconsistencies found in the text had to be reconciled with the inference to specific confines within the parameter of a particular commandment—but, in the end everything had to fit together perfectly. There are nearly a million words embodied in the 24 books of the Written Torah divided into three sections: the Five Books of Moshe, the Books of the Prophets and the Other Writings; together they comprise the **Cud/Cup** filled with the constant merging and fitting together of disparate ideas fit to perfection seamless as water.

THE GREAT ASSEMBLY

The people with the task of codifying the Torah were called, the People of the Great Assembly who gathered together in the Land of Yisroel making sure every letter in the Torah was acceptable to all the Jewish People. The Written Word anchors the Torah; without the accuracy of the Written Word, the Torah would meander into the abyss of conjecture.

Once the Written Word was codified then the debate began as to the possible meanings incorporated within the written text—the beginning of the Oral Torah. Both the Torah and the Temple were being rebuilt after the end of the exile; both required a strong foundation built from stone.

Though the two stones engraved within the Ten Commandments were missing, having been hidden before the destruction of the First

Temple, with the help of H eaven and the last remaining prophets, the verification for each word was possible. What began in the wake of the prophets was a tradition passed down from one generation to the next—the entire knowledge of the Torah. The ultimate game, a challenge to each generation to produce a new vision of the Torah based on the language and events of the time, yet never veering from the written text.

The People of the Great Assembly represented a node in Jewish history from which all subsequent branches must merge; the codification of the written text was paramount to the coherence of the Jewish People. The Jewish People were spread out throughout the Persian Empire, which span from Ethiopia until India; some were lost to assimilation but many remained in what was referred to as Babel—the place of the exile. There were large communities with established places of learning who had accomplished melding with their environment—but, the source of the Torah and the 613 Commandments was eternally in **Yerusahliam/Jerusalem.**

Gone were the days when the curious for adventure would travel to the Land of Yisroel and go to the Temple where ten miracles occurred; seeing one people eating one food and serving God in one way made the sojourner stay and join the Jewish People. Gone too was the ability to remain aloft from the world, acting as a beacon to the Earth—God had spread the people abroad to be individual beacons to the world. The Second Temple would last 420 years before the entirety of the Jewish People was cast to the wind.

The People of the Great Assembly acted as a fulcrum—the hub of the wheel from which the Jewish People would revolve until Moshiach would come and lead the world into the new paradigm. Just as the Jewish People entered into the world through the exile of Babylon so too did the world enter into the Land of Yisroel; the common language was Aramaic, a conglomeration between Arabic and Hebrew spoken throughout the Middle East. Had God brought about the destruction and exile for the purpose of integrating Jewish knowledge into the world?

Besides the codification of the Written Torah, another notable achievement of the People of the Great Assembly was the compilation of a Book of Prayer. Since the Jewish People were thoroughly spread out, coming to the Temple to perform one's obligation was difficult, the wise People of the Great Assembly understood how to make the Temple available to every one through the introduction of set prayer. Through mentioning the obligations of the Temple, each individual is able to participate in the commandment given only to the Kohanim—prayer is in the place of sacrifice.

PRAYER

The commandment of prayer to all human beings is, When you are in need, ask God; prayer can be accomplished in one's heart or in one's mind—a prayer can be spoken, thought or felt and God will understand. The basis of Jewish thought, right from the very beginning was the acknowledgement, God is present in every aspect of life, animating each movement, from the smallest atom until the largest spiritual plane—the proof to this is prayer.

The Temple represented communal prayer which was accomplished through a yearly census requiring each citizen to give a coin worth half a sheckel of silver; through counting the coins, the number of citizens in the population was known. The money collected was used to provide the Kohanim with animals to be sacrificed for all the people—the communal prayer. But, if the people were too far away to come to the Temple, then the Temple needed to be brought to them. The Great Assembly constructed prayer by utilizing and incorporating specific commandments to the Kohanim from the Torah.

Prayer is a sacrifice of one's time and energy, and unlike Torah, which was called by the name, **Sha Olam/Eternal Time**, prayer was termed **Sha Chai/Temporal Time** since prayer should always be new and spontaneous—prayer comes easy and leaves quickly. Torah is engraved into the soul; the more one works at Torah, the more permanent is the engraving. Prayer needs to be new and different, because only then is prayer real. The Great Assembly of elders returning from Babylon composed prayer to be a daily journey to God, through praises from the songs of Dovid and blessings collected from heavenly letters floating down to Earth.

Prayer substitutes for sacrifice; the sacrifices were called, Food for God— just as food satisfies for the moment, so too prayer is a thing of the moment. A reciprocal relationship between man and God is where, we feed God through our prayers and God responds to us through the answer to our prayers. The People of the Great Assembly prepared for God a sumptuous feast, where each person could serve up, in their unique and particular fashion, a tasty dish. The main benefit of prayer is the recognition, through our need and incompleteness we come ever closer to ever-present God by asking for what we need.

Knowing the structure of the heavens, the People of the Great Assembly incorporated the geometric configuration of the spiritual worlds into the structure of prayer to help lift the supplicant beyond the heavens and closer to God. As the Cabala states, *The bringing forth of sacrifice represents the bread brought to the husband—the Seraphim* (fiery angels) *devouring the sacrifice adds to the spirit of life and they lift and raise the Throne with the image of human upon the seat which is on the level of the LightWithoutEnd which is beyond creation. All of this occurred when the* ***Bait HaMikdesh/Temple*** *was standing.*

The basic concept of prayer is the interaction between the human being and the Creator, with the heavens in between; the angels are the spiritual source of the animal supporting the image of a human being on the throne being lifted higher by the influence of the animal. So too, here on Earth, the animal nature of the human being has the capacity to raise the soul closer to God, because the Earth is the place of freedom of choice.

ANIMAL SACRIFICE

The animal is the personification of naked life—able to run, eat and procreate. The animal is neither good nor bad since the animal is instinctual, yet the animal has a soul, an animal soul, longing closeness to the Creator; through the action and intention of the human being, the animal soul is able to ascend into heaven and become closer to God.

Much like the human body, having no implicit built-in morality, just the instinctual actions of its particular nature, can be swayed into dedicated service, because the ultimate desire of the body is to be used as a vessel. The animal lives from the earth's vegetation and from other animals less strong; the principle objective of the animal is to survive. The human being, possessing freedom of choice, is able to elevate the animal and all that the animal lives from by subjugating the animal to further God's plan of creation.

The spirituality of the human soul is far beyond the animal soul, but physically, the opposite applies; the human being is weak and sensitive, unable to withstand the elements— requiring clothes and heated homes. This relationship between animal and human is reflected in the vision of the celestial chair being lifted higher into the heavens; the animal is elevated through means of lifting the chair while the human is directing the path to be followed through the power of freedom of choice and intention of the heart. This collaboration between animal and human is the prototype to understanding the relationship between creation and Creator.

The sacrificing of particular animals required in the Torah are reflected in the natural inclinations of the Jewish People who are like lambs, goats, oxen and turtle doves; these animals touch a primordial core—speaks to the heart. The Temple was a place where prosaic life took on metaphoric meaning. The natural inclination to survive was being mediated by a longing of the soul to come closer to the Creator of Heaven and Earth, no matter what the cost or difficulty. The Temple is a place of life—there was prescription and law but there was also free choice.

The concept of free choice implicitly implies the ability to chose what is wrong; there is a compulsion in the human being to do wrong—the world, and the varied levels of life inhabiting the world, is continually subject to the whim of our desire. The animal is easily swayed by physical pleasure appealing to the senses; without human intervention the animal would merely follow instinct—the human being, who can bring the animal higher or cast the animal into the abyss of the **Sitra Achara/Other Side**, is the ultimate arbitrator between Heaven and Earth.

The Temple is the place to bring the reluctant animal to be sacrificed before God. The Midrash-Book of Metaphor exemplifies this idea with the following parable, Once there was a precious orchard of delicate fruit; the owner of the orchard had to hire watchmen to protect the fruit, but worried, the watchmen themselves were not being watched. He devised an idea; he would hire a blind man who was strong and a lame man who could see. When fruit was missing the blind man said, He did not know who had taken the fruit since after all he was blind. The lame one said, He could not have taken the fruit since he could not walk. The owner put the lame man on the back of the blind man and beat them as one.

The Temple stood as a way to fuse the animal and the human into one unit as is stated in the Torah and explained in the Talmud-Book of Law, *Love God with all your hearts*—meaning the animal and the human merged in love for the Creator.

RITUAL

To make a ritual out of prayer endangered the spontaneity accompanying such requests; only with great wisdom and care could the intention of prayer be preserved within the confine of a text. Rituals were borrowed from the service in the Temple like bathing the hands to come before God in cleanliness and covering the head to approach with humility. The three times during the 24 hour day when animals were sacrificed and burned on the altar became the appropriate time of prayer: morning, afternoon and evening.

The first parameter of prayer was set in the geometric configuration of a bell-curve with the six words of Shema/Listen being at the top; Yisroel; YHVH; **Elohainu/Our God**; YHVH; **Echud/One** being at the summit. The long, arduous path up the hill of prayer was laden with song and story from the history of the Jewish People; these beginning pronouncements and adorations were to be delivered loudly with all the energy the animal soul. Sacrifices happened amid the songs sung by the Tribe of Levi before an orchestra of varied instruments of an allotted number—but violins could be without end.

The down side of the hill brought the intention of prayer down to the reality of world with the admonition to, *Know this day and put this day into your heart, YHVH is Elohim in the heavens from above and on the earth below—there is no more;* and the hope, *To quickly see the beauty of Your strength wipe away the excrement from the Earth and cut down the idols, establishing the Kingdom of Shadi.* Prayer ends with the prophecy, *And God will be King on all the Earth; in that day God and the Name will be One.*

Between the proclamations of One are the praises and the prayers; the joyful supplications and sorrowful recriminations fluctuating in song culminating in the complete abandonment to the One. The moment the animal is slain, the moment when prayer becomes more important than body, the animal leaves temporal existence and returns back to the Oneness of God. After entering into the realm of One, everything is downhill from there—bringing the light of the Oneness into the world.

The beginning of prayer acknowledges the Creator created Heaven and the Earth; this first phase is representative of the lowest of the four spirit world, **Asiya/Action.** The next part of prayer corresponds to the second spirit world known as **Yetzirah/Form**; here is where acknowledgement turns into praise. The third part of prayer leads up to the Shema Yisroel by recounting the ways of the angels—the source of the animal. The fourth and final spirit world **Atzilut/Close** is performed by standing at attention while pronouncing the 18 benedictions corresponding to the spine of the body.

Incorporated also into prayer were the commandments to adorn one's self with Talit and Tefillin. The Talit is a four cornered garment affixed with eight strings tied with five knots on each of the four corners called, Tzitzith whose gematria is 600; when added the number of the eight strings tied in five knots produces the number 613; the Tefillin are black leather boxes incasing a scroll of the Shema Yisroel to be worn on the head and arm. Both commandments, Talit and Tefillin are ways to remember and be subjugated to the 613 Commandments.

LITURGY

The concept and configuration of prayer, much like the bones of a skeleton, had to be dressed in flesh. Much of the text is drawn from **Tehilim/Songs** written by Dovid exalting creation; speaking directly to God, the Tehilim pleads and glorifies while meandering through the many stages and situations life presents. Chana, the mother of Shmuel the first prophet who anointed Dovid to be king is, as the Talmud states, the mother of prayer—then surely Dovid is the father of prayer.

The People of the Great Assembly were not beyond cutting and pasting and sometimes even changing the end to suit their necessity, but every word was debated until there was consensus. There were times, as recounted in the Talmud-Book of Law, when the entire congregation was perplexed and could not come to resolution; a paper would float down from the heavens with words inscribed in the celestial language to be spoken upon the Earth. Many of the 24 books of the Torah were utilized putting together the liturgy of prayers strung together phrases or excerpt pieces like, **Shir HaYom/Song of the Sea** sang while crossing through the sea, split open.

By appealing to the extraordinary history of the Jewish People, the prayers acted as a way to awaken, through the mention of ancestors and their deeds, an experiential spontaneous outpouring of emotions. Also, incorporated into prayer was the reading of the Torah three times a week on Monday, Thursday and the Shabbot. Moshe had divided the Torah into weekly portions, each divided into seven parts so the entire Torah (The Five Books of Moshe) could be read to completion each year week by week. The Jewish news of each day is the part of the weekly portion being read throughout the Jewish World—through the Torah, the people achieved oneness and unity.

The pinnacle of prayer is the six words of **Shema Yisroel/ Listen Yisroel** YHVH Elohanu YHVH **Echud/One** which is divided into three proclamations: connection through the lineage of Yaakov (**Shema Yisroel/Listen Yisroel**); nature hides God (YHVH Elohanu); and YHVH is One—omnipresent, all the time, to everyone. This commandment requires meditation upon these six words through the reading and the physical articulation of each word; the way in which this commandment is performed is by placing the hand over closed eyes then reading these six words engraved within the mind.

Inside the Tefillin worn on the head is written the Shema Yisroel (the six words and three paragraphs that follow) on four little scrolls divided into four compartments representing the two lobes of the brain and the mirror image of knowing; inside the Tefillin of the arm, in one compartment tied close to the heart again is written, the Shema Yisroel. The human being, a conglomerate of man and animal, requires guidance to both the head and the heart; the head is compartmentalized, whereas the heart feels everything. Saying the Shema wearing the Tefillin strapped to head and heart while wrapped in the Talit, is the pinnacle of enacted prayer.

The Talit, with the four Tzitzith at the corners representing the four directions, and the Tefillin of the head and heart representing up and down—the six direction connected to the One like the six days of the week connect to the Shabbot. The Talmud-Book of Law explains, The concept of One is encoded in the three numbers comprising the word **Echud/One**: **Aleph/1**—the Singularity of God; **Chet/8—**the seven heavens and the Earth**; Dalet/4**—the four directions.

In the Written Torah, the Shema Yisroel is written with a large letter Aiyin and a large Dalet making the word, **Aide/Witness**. The human witness, at this lowest level of creation, brings God closer so we can whisper our prayers and explain our needs.

ONE

The concept of One, in modern terms, is referred to as a Singularity, an unfathomable idea in a creation full of duality and multiplicity—all creation divides with the exception of One. Aleph/1, the first letter of the word **Echud/One**, has in addition a simple meaning, Leader; Aleph read backwards renders the word, **Peleh/Wondrous.** The answer to the dilemma of contradiction and other conundrums is found in the concept of One as portrayed in the six words: Shema Yisroel YHVH Elohanu YHVH **Echud/One**—Hear Yisroel; YHVH is Our God; YHVH is One.

One of the interesting nuances in the word **Echud/One**, if the phrase was trying to convey the unity of God Torah would have said, Yuchid/Unity; but, **Echud/One** refers to the wondrous way the Creator can be both in the world at every level and also be completely outside of creation. The six words of the Shema Yisroel are not just the pinnacle of prayer, but also the last words uttered on the dying lips of the Jewish People—an acknowledgement that wherever dispatched, God will be there.

Avraham is widely credited for having brought monotheism into the world, but this is untrue; neither Avraham nor the Jewish People were overly concerned with how many gods people prayed to, since there was common knowledge throughout the world of the God of gods—the Creator. Equally common was the assertion, God created creation then left the workings into the hands of the stars and constellations. In very simple ways, Avraham taught the world to pray to the Creator of the Heavens and Earth in their heart, in their mind and on their lips— acknowledgment to the eternal presence of the Creator.

How the One is able to communicate to the many is the secret of creation. Avraham was first to bring this concept to the Earth and wrote in his book the **Sefer Yetzirah/Book of Form**, the explanation and spiritual meaning of the 22 Hebrew letters—**Lashon HaKodesh/ Holy Tongue**; later, the concept of **Echud/One** and the relationship with the 613 **Mitzvot/ Commandments** was expounded upon by the People of the Great Assembly who divided the teachings of the

Torah into 600 categories—13 is the gematria of the three letters comprising, **Echud/One.**

The People of the Great Assembly besides establishing prayer and framing Jewish Law, also admonish the future by insisting on three things: patience before judgement, establishing many students and making a fence around the Torah. Up until the end of the First Temple, the Kohanim were the repository of Torah, though there were great scholars throughout the tribes, but the People of the Great Assembly were a conglomeration of what remained of the Twelve Tribes including Levi and Kohan; the time of the Rabbis begin with the return to the Land of Yisroel from Babylon-Persia—with them they brought the Written Torah and the Oral Tradition.

In the five hundred year span, the last half of the thousand years of **Netzak/Victory**, is the time of the Rabbis, who taught us how to interface the knowledge of the Torah with the knowledge of the nations. The greatest of those Rabbis, Rabbi Meir, the assumed author of the **Mishna/Learning**, seminal work of the Talmud-Book of Law, if not otherwise noted; and Rabbi Akiva of whom the Talmud says, What was not given to Moshe to understand was given to Rabbi Akiva—descendant from converts.

LOVE

In Hebrew the word for love, Ahev, (Aleph-Hey-Bet) is made up of two factors: Aleph representing *I* and Hev (Hey-Bet) meaning, To Give—I Give. Immediately following the Shema Yisroel is recited, *Love God with all your heart, with all your soul and with all your might.*

The word for heart, Lev (Lamed-Bet), is written instead with two Bets indicating the need to love God both with the animal soul and the human soul.

The heart, and soul and the might indicates, one must love God with the heart connected to the Creator, their mind, where the soul interfaces with the brain; and **Moad/Very** whose letters can be rearranged to read Adom—loving God must be beyond instinct; the way of the human being to always strive for more is indigenous to our specie. Three hundred years ago the Baal Shem Tov was asked: how to love God? The Bal Shem Tov replied, *Love what God loves; God loves creation.*

The implicit meaning of the word **Ahev/Love** as I Give indicates the kind of relationship the Creator has forged with the human being—one of giving. God gives to us life, sustenance and incident; by performing commandments, we give back what God wants. To the world God gave seven commandments: not serving other gods, not profaning the Name, not spilling blood, not stealing, establishing laws of sexuality, establishing laws limiting the use and treatment of animals and the environment, establishing courts of law. To the Jewish People God gave 613 Commandments.

The human body of seven limbs (head, torso, arms, legs and sex) populating the Earth on seven continents on the seventh planet in from Pluto at the rim of our solar system. The limbs of the body are ascribed in 248 divisions connected by 365 veins or bridges as displayed in our own solar system where the seven outer planets are bounded by the 365 day revolution of the Earth and the 248 year revolution of Pluto. Similarly, the 613 Commandments given to the Jewish People are divided between the 248 Positive Commandments and the 365 Negative Commandments.

The implicit message inscribed in the configuration and geometric coincidences of creation describes a creation built on love—God loves creation and creation loves God. The seminal first book of the Torah **Bereshet/In the Beginning** are stories about love relationships between man and wife, strife between brothers, the righteous against the wicked, the barren and the miracle—all types of relationships are explored because each one teaches us how to interface with the Creator and how to love. Life is a contest of love. Love is the basis of life, but to maintain the needed balance, to be always in love, is a battle.

After return from Babylon and finding the land in disarray, the Temple Mount bereft of body and soul, the Jewish People felt as divorcee returning to her past and finding her home demolished and the land gone to seed. Like a woman alone, the Jewish People brought the land back to life, nursing her back into health—building the Second Temple. The Second Temple was a modest structure compared to the First Temple living in the generational memory of the past; the Temple gave the experience back to the people—even though the soul was gone (the Ten Commandments were missing) still, even just a replica, her power was great.

Chapter Six

Beauty

BUILDING

No longer in her youth, a small portion of the Jewish People returned to the desolate place where once the land swelled with reverie; abundance, fertility and miracle was all memory—a divorced woman retuning back to her old neighborhood brick by brick rebuilding her old house. The plenty of the past and the pleasantries of life were gone; the world had entered into a new time— the victorious triumphs had been ensnared in defeat and a new type of victory became predominate.

Half way through the 1000 years of **Netzach/Victory the B'nei Yisroel/Children of Yisroel** found themselves a defeated people; broken dismantled and dispersed even the reclaiming of the land was not without controversy. After Cyrus the Great defeated Babylon he let 40,000 Jewish People return to Yisroel, but it was not until many years later that the People of the Great Assembly returned to rebuild the Temple, as a result the people had become divided as to the meaning of their heritage.

Just to reclaim the land and live as other nations lived was enough for many, but the zealots wanted to return the former glory; the People of the Great Assembly understood what had been written in the book of Lamentations where it exclaims: Baka TaBaka/Crying and Crying— two Temples will be built and two will be destroyed. They understood the concept of **Netzach/ Victory** not in its conventional meaning, but rather as a pathway to Moshiach. The sign was clearly written in the Torah concerning the three children of Yehuda: Aire/Awake (the First Temple), Onan/Bereft (Second Temple) and Shiloh—the Third Temple which would be built in the time of Moshiach and last forever.

Defeat and destruction is part of **Netzach/Victory**. War, like life, goes through phases; gains are balanced by loses and winning is a subjective call; the last victory is when the war is over. Likened to birth, the travail to bring Moshiach into the world is commensurate with war where the ultimate victory is life. While the fetus swims in the embryonic fluid being fed through the stomach with eyes closed and consciousness shut off, the child can not comprehend birth and life, but once born into the air the child is forever awakened.

The three Temples are compared to the three trimesters of pregnancy: the First Temple is the beginning of life and the Second Temple is the intermediate step before the last trimester culminating in birth. According to the Talmud-Book of Law birth is possible from the beginning of the third trimester; Moshe was born after six months and one day whereas Eliohu, a later prophet, was born after 12 months in the womb. Eliohu had a body so refined because of the length of time spent in the womb that upon death he rose up into heaven with his body.

The nine hundred years between the building of the First Temple which stood for 410 years and the building of the Second Temple built after 70 years of exile stood for 420 years until burnt to the ground by Rome—**Ashun/Smoke** is gematria 420, but now the travail of the expected birth has stalled for 2000 years. It was widely expected that the 1000 years of Netzach would end with the birth of Moshiach and the building of the Third Temple, but that's not what happened— something else happened instead.

BLESSING

Two of the decrees instituted by the People of the Great Assembly indicates the state of the Jewish People during that time: one, the Yovel—a fifty year interval that saw all lands returned to their original owner—was halted since only a small minority of the people returned; and two, the requirement from the Rabbis to make blessings within the Temple.

In the First Temple blessings were not required since, as the cabalists explain, the First Temple was called Mother because the spiritual level introduced to the world was akin to the place from where all light emanates; compared to the Moon reflecting the light of the Sun, **Shechina/Dwelling** is mother of creation. The purpose of a blessing is to draw down; witnessed by the etymology of **Bracha/Blessed** with the word **Brakim/Knees**—when we bend our knees we become closer to the Earth.

At the level of **Shechina/Dwelling** the requirement was to open up and awake, but the Second Temple called Daughter was related to the tenth of the Ten Sefirot called **Malchut/ Royalty**. All blessing are prefaced with the six words: **Baruch/Bless** (draw down)—**Ata/You**—YHVH—**Elohanu/Our God—Meleck/King—HaOlom/The World** as a way of attenuation of the light through a cord connected to heaven that bends, ripples and pull from the source of all blessings; the uttering of these six words help draw the light from the interior of God's essence (You) through the Name YHVH revealed through nature through the Name **Elohim/God of Plurality** down to the last of the Ten Sefirot (**Malchut/Royalty**) to the earth (**HaOlom/The Earth**)

The purpose of the Second Temple was to draw the light down closer to the eEarth where it would be miraculous. During the First Temple miracle was not of importance, but rather the subtle difference between worshipping God and worshipping something else; that sustenance came through the stars was obvious, but to steal from the stars without acknowledging the Creator of Heaven and Earth was blasphemy. During the time of the Second Temple the miracle was a sign, but again came the ambiguity whether the sign was coming from the heavens or was it nature being manipulated by subtle forces here on Earth?

The institutionalization of blessing was a clear sign from the People of the Great Assembly that the purpose of the Second Temple was to draw down the Third Temple in her place. The Third Temple was destined to come from the heavens made of elements indestructible upon the earth—a beacon to the world declaring that Moshiach has brought peace to the Earth. There were some even amid the scholars who scoffed at the expectation.

Rabbi Yochanon was once lecturing about how the pillars of the Third Temple will be made out of diamonds when one of his students asked: how is it possible a pillar can be made out diamonds when we do not find a diamond in all the Earth larger than a fist? Rabbi Yochanon did not answer. The student later was on the sea traveling on a boat when the skies opened up and he viewed the angels chipping away at the pillars of the Third Temple.

He returned to his teacher, Rabbi Yochanon, and announced that he had seen with his own eyes the diamond pillars of the Third Temple. Rabbi Yochanon rebuked his student for not trusting in the words of the his teacher and the student "turned to dust"—reverted back to his essential being without self imposed suppositions.

SHECHINA

The human being's ability to speak is closely related to the aspect of God known as **Shechina/ Dwelling**. Also known as Daughter she is represented as the tenth of the Ten Sefirot called **Malchut/ Royalty**—a king rules through his word in decrees sealed with the royal stamp. But, the voice of royalty is patently feminine. Why God chose to be revealed in the feminine is because the quality of feminine was to hold within and released through **Tzimtzumim/ Contractions.**

Speech is a process of minute contractions conducted across the five parts of the mouth: lips, tongue, pallet, teeth and throat; the five constrictions of the mouth is related to the letter **Hey/5** the least articulated letter and the breath is related to the letter **Yud/10**—the Yud, the smallest of the 22 Hebrew letters, is just a dot. The Yud-Hey, the first two letters of the YHVH, is indicative of how God used the concept of male and female to create creation through speech.

The Zohar-Book of Secret says in the beginning Avraham mistakenly thought that God was female because he did not see the masculine hiding beyond. Similarly, a child at first is much closer to the mother and does not recognize the father's contribution immediately; the cabalists later put the Yud inside of the Hey looking like a stone inside of a house—the Ten Commandments within the **Beit HaMikdesh/Temple**. In the Second Temple the Ten Commandment with their five attributions—the stone, ark, the Staves and the Two Cherubim— were missing and God no longer spoke to the people.

Nonetheless, the People of the Great Assembly knew that building the Second Temple would allow the Kohan Gadol to enter the empty chamber called **Kodesh HaKodeshim/Holy of Holies** and communicate through a labyrinth of similar structures engraved into the firmament until the world of Atzilut where the Shechina merges with our world. In each of the Four Worlds there is an ephemeral Temple where the words of the Shechina vibrate from between the outstretched wings of the Two Cherubim and the Two Staves like breasts pushed out against the veil.

The Talmud-Book of Law says that if the Kohan Gadol has one stray thought while within the **Kodesh HaKodeshim/Holy of Holies** he would immediately die. When the Kohan Gadol returned alive he would utter the YHVH and the Temple stones would shrink and those who were pushed together from the multitude entering the Temple would suddenly find ample room to fall on their face and prostrate themselves on the expanded stone floor; yet, when they left the Temple none could remember the pronunciation of the YHVH. Though the Second Temple was bereft of the Yud-Hey in the form of the Ten Commandments and the five attributions, still the Vav-Hey, the latter two letters of the YHVH which is related to emotions and sexuality; brimming with the cacophony of noise produced from the animals being slaughtered amid the constant streams of endless people come to pray—to see and be to seen— was the choir of Leviim playing countless instruments and singing the praises as written down by Dovid the first from the line of kings from the Tribe of Yehuda-—Jewish.

BEIT HAMIKDESH

The construction of the Temple in the form of a human being was a body, the Ten Commandments engraved into a cube of stone symbolized the soul of the Temple, but the Kohanim and Leviim were the heart of the Temple. The heart has two aspects: receiving the old bad blood and releasing new rejuvenated blood. The Kohanim were the blood carriers and the Leviim gave breath, lifting their voices in song and watched at night from the towers of the Temple. The night watches were divided according to the three parts of the night, but they watched for the honor of God not because of fear. The Talmud-Book of Law says occasionally at night a scream would be heard of a sleeping Levi being harshly awoken in the middle of his watch by an elder.

The progenitors of the Twelve Tribes, the twelve sons of Yaakov, were actually thirteen since the two sons of Yoseph were each designated a tribe because of the statue their father enjoyed in Mitzraim was considered first born thus receiving double portions. After the incident concerning the Golden Calf the tribe of Levi was chosen by God to lead the people and was separated out during the census count. The tribe of Levi was prohibited from ownership in the land and Kohan, a lineage stemming from Moshe's brother Aaron, a family within the tribe of Levi, was to be wholly destroyed in the service of the people.

It was the Kohan who could judge between pure and impure; it was Kohan who guided the officiating of the sacrifice and it was the Kohan who received gifts from the people in appreciation of their work. With alacrity and precision the Kohan did his work; he was exact because a false step in the Temple was tantamount with forfeiting one's life. During the Holiday of Pesach (Passover) there were so many sacrifices that the blood required to be pour at the base of Altar had to be transferred through cone-shaped silver vessels that could not be set down least the blood coagulate—the Kohanim were able to line up and pass full cups of blood one way while returning empty cups the other way and never dropped a cup or spilt a drop.

The cabalists explain that the three branches of Levi: **Kahot/Gathering; Gershon/Cast Out;** and **Marari/Bitter** indicate the responsibility, like the left side of the heart, to gather the blood and cast out the bitter part while Kohan, the right side of the heart, injects kindness into the beginning of the new cycle. This convergence of kindness and severity is the key to life—the sinusoidal wave that vibrates through all things.

The Temple is in the configuration a human body—the head being the Holy of Holies where the Ten Commandments are kept in a cube beyond time and space, and the Torso the Hakel/Hall where the Golden Candelabra gave light, the Golden Table provided bread and the Golden Altar where the incense was burned; the lower body, the altar made of stone for the sacrifice of animals were consumed by the fire which fell from the heavens in the form of a lion—was brought to life by the tribe of Levi who would raise their voice in exaltation to the Almighty with an orchestra of limitless violins; the people danced and rejoiced.

SANHEDRIN

Abutting the Temple ground was erected a great hall where the **Great Senhedrin/71 Elders** sat and decided the law. Unlike the austere setting of modern day Western Civilization, in the great hall next to the Temple because of the height of the ceiling, many conversations could be conducted simultaneously. Those cases which could not be settled came before The Great Sanhedrin.

The judgment of the rabbis needed to be close to the Shechina in the same way the heart was divided into two sections. The smaller Sanhedrin made up of smaller courts from different towns would adjudicate the less difficult cases; after coming to the Temple, after seeing and being seen, one wanted justice and at the foot of the Temple is where they found it.

The smaller Sanhedrin was presided over by the **Nasi/Prince** who was the interface between the small and the Great Sanhedrin; the Great Sanhedrin was officiated over by the Av **Bet Din/Father of Judgment**—these two leaders were the die from which the Rabbis were cast. Not only was there two leader, one that presided over the people and the other over the Rabbis, but the great hall at the foot of the Temple was precariously place on the boundary of the Temple grounds so half was within and half was without.

The Talmud-Book of Law condenses the collective wisdom of the People of the Great Assembly into three things: be careful with severity; make many students; and make a fence to protect the Torah. These three injunctions summarized hundreds of years of renewal to reconstitute the Jewish People. Severity requires fences needed to be built and many disciples developed. The pristine memories of the First Temple when the Jewish People were pure and innocent had been overtaken by the struggle for survival—the Torah had not changed but the rules of life had changed.

It took a few hundred years for the People of the Great Assembly to disperse leaving an institution in it place and it was from the people of the Great Assembly that the Rabbis emerged; the greatest among them was Shimon **HaTzadik/The Righteous** who put the three statements of the People of the Great Assembly into context by saying: the world stands on three things— Torah, Service and Giving. The subtle distinction that Shimon **HaTzadik/The Righteous** gave to the ideas of the People of the Great Assembly began the rabbinic transmission.

Where once twelve tribes worked the land and performed the Commandments while the Tribe of Levi was dedicated to service, now a new institution had been formed based in the study of the Torah. It had been taught that the First Temple was destroyed because the people, though they did the Commandments did not love the Torah. The Rabbis took it upon themselves to make sure that no one would trample upon the Torah; by building fences strung with endless laws about minute details, the fence became a wall. The general rule was: you could not be stricter than the Rabbis—to do so with be an error.

RABBIS

The Rabbis were known to one another as **Chaverim/Friends** because the acceptance of divergent views was incumbent upon them all; the famous arguments and disputes which rose up amongst them was a function of this requirement to see the Torah from the eyes of their friends. Rabbi Akiva said: the Commandment, loving your friend like yourself is the basis of the Torah.

What might seem to be a formula for chaos—if everyone's view is accepted then where is the unanimity? But, the Rabbis were not arguing about right or wrong, rather they were trying to punch holes in the armament of their friend. Each of the 613 **Mitzvot/Commandments** which they discussed was a piece of the armor made strong, forged into metal through the heat of these disputes.

An errant idea was quickly repudiated—unique standpoints were encouraged but not at the expense to the coherence of the Torah.

The 613 **Mitzvot/Commandments** a combination of the 248 positive commandments and the 365 negative commandments are representative of the 248 limbs of the human body and the 365 connecting vessels—the performance of these commandments strengthen the metal of the armor. The admonition to make a fence around the Torah began with the armor of one's views. Most people conducted themselves according to the Torah in the way in which the family as a whole within a particular tribe understood their requirements in the Torah, but the Rabbis each developed their own unique way of viewing the Torah as a whole without contradiction— this was the real love of the Torah.

In their discourses with one another they would learn how to refrain from judgment since what appeared at first to be completely contradictory often turned out to be just as logical when viewed from another perspective—the essential multi-dimensional truth of the Torah was being revealed. As might be expected great teachers arose with mighty minds that challenged the masses that came to hear their discourse. Each teacher became a school and the great teachers formed academies that encompassed vast amounts of teachers. In this way the institution of the Rabbis grew.

There were many things the Rabbis held in common but it was the subtle differences which became huge chasms since the smallest deviation, like an intricate mathematical method, could throw the entirety into disarray. Perhaps to the outsider this dance of ideas seems like pointless confusion over minuscule divergences, but the truth, for those **Chavarim/Friends**, this was the most engaging of sports pitting agile minds against rock-hard truths.

Above them all was the **Rosh HaDor/Generational Head**—the one man who somehow personified the entirety of his generation. The Talmud-Book of Law traces those generations by their heads and what seminal ideas were attributed to them. Shimon **HaTzadik/The Righteous**, the last remaining remnant from the People of the Great Assembly, challenged the Jewish People by reminding them the world stood on three things: Torah, Service and Goodness. The head of the generation could see furthest into the future and tried to explain that while the Rabbis had to protect the Torah, it was more important for the people to live the Torah.

THE PEOPLE

The people who populated the Land of Yisroel during the time of the Second Temple were diverse since the Shmita, the fifty year interval that returns the land back to their original owners since the time of Moshe, had been abolished by the Rabbis. It is noted by the cabalists that the seventy souls that descended into Mitzraim with Yaakov and his 12 sons entered the Land of Yisroel hundreds of years later with seventy families who divided the land between the twelve tribes.

But, those archaic divisions had been erased with the destruction of the Temple and the subsequent exile into Babylon; as the old boundaries eroded, new division began—ensuing struggles occurred, disputes developed and a new society sprang up in a constant redefining of the Jewish People. The people had been warned that the destiny of the world depended on how the Torah and the service of the Temple would be given to the world.

Having lost the ability and the mental agility to pursue the Torah, the people became reliant on the Rabbis as communities began from the Jordan River to the sea; from Lebanon in the North and Gaza in the South different schools of thought emerged. Not all of the people resided in the Land of Yisroel, many remained in Babylon where communities and schools of equal stature persisted; the people knew the teachers from Babylon as Rav and the teachers inside the boundaries of Yisroel as Rebbe.

Because life was not as easy and plentiful as it had been during the 410 years of the First Temple and because the Jewish experience was rapidly changing, the Torah was finding new application. The Torah, known as the Tree of Life had been planted in the Land of Yisroel. The Temple and the Sanhedrin was the trunk of the tree that branched out and bore fruit throughout the land rooted deep with the 24 books of the Written Torah.

It says in the Talmud-Book of Law comparing a person to a tree: one must be limber in the wind as not to break; the winds of conquest were blowing through the Middle East, Persia had fallen to a brash young man named Alexander who was headed for Yisroel. What would this mean for the Jewish People? Alexander, who carried with him into battle the book of Homer, was uniting countries through the amalgam of cultures.

When Alexander came to Yisroel he had a dream in which the distinctive face of an old man looked at him; on a mountaintop not far from Yerushaliam that old man stood and waited— it was Shimon **HaTzadik/The Righteous**. The old man had come to plead for the sake of the Jewish People that the king of Greece should not put his effigy in the Temple. Alexander retorted that he wanted to be remembered. **HaTzadik/The Righteous** made a deal with the king: the Jewish people will name their children Alexander—a custom that continues until today— and the king did not disgrace the Temple.

GREECE

Greece was the next step into the exile; though a remnant of the people return to the Land of Yisroel, still the darkness of exile continued its descent. Like a plague of locust vast armies invaded and stayed bringing with them various cultural variations. The Jewish people were use to living with the nomadic Arabs who trekked across deserts carrying spices and other merchandise. Foreign soldiers were another influence, but it was the way of Alexander to encourage his troops to settle down in the countries they invaded so that the Greek language and customs would be predominant. It was called Hellenization.

The Greeks took a humanistic view of the world; beauty was tantamount and truth was relative. The human body was glorified and nudity was explicitly allowed. Sexuality was a vehicle of Pleasure and life was just for the living. In dramatic contrast to the modest ways of the Jewish and Arab People, Greek culture was decadent and egotistical. The gods they prayed to were merely glorified representations of themselves.

But, the Greeks were purveyors of knowledge and had collected a variety of wide- ranging artifacts and ancient wisdom which they openly usurped as their own. According to their humanistic philosophy the melding of cultures would bring about a greater human being where the strong and the beautiful rule the Earth. It is an easy philosophy to sell to the unsuspecting and through it they were planning to conquer the world, but unfortunately Alexander died in his early thirties leaving in his wake an amorphous vision of the future.

In the Land of Yisroel, the Hellenic dream was a festering sore. The Jewish People led a spiritual life where beauty was seen in modesty exemplified by Moshe who looked deep within. The beauty beheld with the eyes was inferior to the beauty within; the inner beauty was indicative of the soul enlivened by a ray of light from the LightWithoutEnd that exuded from the Creator of the Heavens and the Earth.

So resistant were the Jewish People to the Greek way of life, that open hostilities erupted, but not into violence; the disparity between the soul and the body took on real dimension in the Land of Yisroel fertile with Truth under siege by the pestilences brought by Hellenism. The dialogue which resulted deeply influenced each people, but the struggle for domination overshadowed their peaceful coexistence. Each, the body and the soul, was determined to define and regulate the movements of the limbs.

The Greeks wanted to conquer the world and the Jewish People wanted to bring revelation to the Earth—through the Torah. The Greeks were not opposed to the Torah or the God of Yisroel—their attitude was: you take our gods and we will take your God. Each wanted to dictate the terms of engagement until finally dictates came down from the Greek governor demanding Torah be prohibited. In addition the Governor suggested that the daughter of the **Kohan Gadol/Head Cohen** be sent to him.

REBELLION

Her name was Yehudit and she was beautiful as was her father as were all the Kohanim who worked in the Temple; it was one of the requirements—only the most beautiful amongst the Kohanim were allowed to actually work within the Temple and the daughter of the head Kohan certainly was a woman of great beauty, intelligence and skill. She took up the governor's offer and went to his palace where she fed him wine and cheese until he became drowsy.

When she emerged back into the street with tyrant's head severed from his body, it was a sign of revolution; though small in number, led by a Kohanic tribe called Chusmonoim latter known as **Maccabee/Hitters**, this small group of valiant troops managed to defeat the great Greek empire and force them from the land. In the wake of their retreat they left the Jewish People a message: they removed the seals from the containers holding the oil used in the Temple to light the candles.

There were other desecrations that took place within the Temple, but none were as explicit as the incident with the oil. The oil used in the Temple to light the seven branched Golden Candelabra was made by gathering the first drop of oil from each olive; this process was supervised by the Kohan **Gadol/Great** who gave his official stamp of approval in the form of his seal. By breaking the seal, the Greeks had challenged the Kohanim to break the law.

It would take eight days to process new oil, but miraculously one cruse of oil had been found with the seal unbroken. The question was whether to use the one cruse to light the candles one complete night or divide the oil so each of the eight nights would have a little light—the later choice was the most logical, but being zealots they chose the former and burned all the oil that night. The miracle of Chanukah was the light continued to burn for eight nights until new kosher oil could be produced.

The Greeks had tried to dilute the Torah, but the Jewish People had appealed to God who helped the Jewish People be victorious and then did a miracle as if to say: do not be afraid to go into the darkness with a little candle because a small light pushes away a lot of darkness. The Jewish People were meant to be a nation of Kohanim to the world—meticulously helping the world maintain life and equilibrium. However, the message could be interpreted in different ways.

The Chasmonean-Maccabeean took the message that their leadership was being acknowledged and extended their reach to include the crown of the king. The king had always been from the tribe of Yehuda and now a Kohan was king. The Rabbis vigorously objected, but to no avail. Once the law had been breached the integrity of the Temple began to erode; what the Greeks had tried to accomplish the Jewish People did to themselves.

The Greeks were leaving, but the Romans were coming.

TZADOKI

Antigonous replaced **Shimon HaTzadik/The Righteous** as the head of the generation; Antigonous taught the people to serve God as a servant who receives no reward. As head of the generation he was trying to invoke in the people a level of love beyond what was natural; the cabalists call it: **Ahavot Tainugim/Love of Pleasure** since Pleasure was the animating influence in the Ten Sefirot. Pleasure is considered to be the highest level within creation; Antigonous tried to show the people how to attain, through their love, a relationship with God at the highest level beyond Pleasure+.

Because of his elevated goal he failed to see the negative impact of his lofty aspirations. People for generations afterwards questioned whether indeed there is reward for the service and requirements dictated in the Torah; the people of the world were obliged to perform seven commandments while the Jewish People, the Chosen People+ were obligated to 613—and there was no reward?

The Midrash-Book of Metaphor relates this self-professed doubt became a school of thought introduced by one of Antigonous' own students whose name was Tzadoki—progenitor of a vast family whose secret aim was to radically change the definition of the Jewish People. Up until now, the Torah defined the substance and place of the Jewish People within the timeframe of the Creator who indicated through the Six Days of Creation and the Seventh Day of Rest that life on Earth would be a seven thousand year journey—the half-way point had been reached.

The Tzadoki rejected the oral aspect of the Torah which the Rabbis had championed; they called the Rabbis the **Perushim/ Explainers** because they were constantly trying to explain the perfection of the Torah with imperfect words—they spoke Aramaic. The Tzadoki adopted Greek as their language and professed melding Yisroel into the world by establishing alliances and forging unbreakable bonds.

The Greek influence had divided the people through the doubt they incurred; people were becoming more materialistic and ceremony was replacing intention. The perfidy persisted and eventually found its way into the Temple where it was rumored that Kohanim were paying for the privilege of entering into the Holy of Holies on **Yom HaKippur/Day of Atonement.** Things became so charged, according to the Talmud-Book of Law, that it occurred once during the holiday of Succot when the men were gathered holding their Lulav (palm branch) and Etrog (a small citrus fruit) as prescribed in the Law that an unfortunate man misspoke his words and incurred the wrath of his brethren; thinking he was a Tzadoki they threw their Etrogim at him thinking he had come to the Temple to make trouble and he died.

The chasm which was opening beneath the feet of the Jewish People was fathomless; the Rabbis hated the Tzadoki and the Tzadoki hated the Rabbis and mistrust was planted among the people. Rome had devoured the Greek Empire and now they were in charge; Rome lived by the sword. As the thousand-year period that represents Netzach was drawing to a close, the prophetic message from the blessings of Yitzchak: when Yisroel falls Esau will arise, was taking place.

The Zohar-Book of Secret says when Shlomo married the Queen of Sheba from Egypt, the Angel Gabriel came down and put a staff into the sand and that was the beginning of Rome; for a thousand years Rome had been percolating from the seeds of Esau and his many offspring including Amalek—enemy to the Jewish People from the beginning. It was into this fray that Jesus was born.

Chapter Seven

Victory

JESUS

The truth about Jesus is found in the mystery of his birth and in the culmination of his death, but the many questions concerning his life are much less relevant. The legacy of Jesus is greatly misunderstood due largely to the lack of any historical evidence substantiating the purported story, whatsoever. It seems miraculous how one man could, in recent historical past, make such an impact on the world without leaving any trace of having been on the Earth.

It was an ironic life he led, one full of misunderstanding and escalating misjudgments. The prophecy of his life and the challenge that he presented to the Jewish People was not clearly evident and everyone had a different take on what was going down. Prophecy had ceased with the People of the Great Assembly—those chosen to transmit exact messages from the heavens. The prophecy of Jesus was not i what he said, for little is known of what he said—all is encompassed in his birth and in his death.

The only trusted sources for piecing together the story of Jesus is in the Talmud-Book of Law and the Koran written down 600 years later. The Talmud is a compilation of stories and law handed down throughout the generations by the Rabbis, but the Koran is prophecy. The Talmud tells the story from the side of the Rabbis and the Koran tells the story from the side of the heavens—both Heaven and Earth were witness to what happened.

The Talmud was put together over centuries of argumentation written down at the end of the fourth millennium, the thousand years of **Netzach/Victory** about 300 years after the destruction of the Second Temple. The Talmud is like a rock in the river of the Oral Law that suddenly exposed in a frothy turbulent whirl of water—different disciplines and diverse ideas. The oneness of water can break up into vaporous spirals and illuminated rainbows with loud crashing voice like a swimmer taking a breath before returning to the oneness of the water. The Talmud was just the place in the river of time where the teachings were exposed.

Like a river, the generations blended one into another through the subtle changes in perception of the surrounding world adapted through the Oral Torah; it was a game of perfection played by the Rabbis under the strictest of rules to preserve the correctness of the Torah. Many views were born from this process of vigorous refutation by the Rabbis whose minds were able to process hundreds of thousands of words until their own opinions were known as **Divri Sofrim/Counted Words** since they were simultaneously conscious of the numerical value of all their words based on the prophecy of the Torah.

The Koran is the new prophecy given to the Arab People because the Jewish People did not accept the prophecy of Jesus; the Koran comes to clarify the truth, but the words of prophecy are not always clear and can be easily misconstrued. Nonetheless, every word of prophecy must be true or the entire prophecy loses its credibility. The Koran is adamant that the Rabbis were wrong and did not take the sign from the prophecy of Jesus.

The sign the Koran refers to is the birth of Jesus through his virgin mother. It is an anatomical fact that virgin girls can, because of the smallness of the sperm, conceive a baby without having had intercourse—this is usually considered a tragedy and certainly not a miracle. What is a miracle is a woman already married with children who, at the time of birth, becomes a virgin—but the Rabbis did not believe the report of the midwives and would not look for themselves out of modesty.

The Rabbis with their encyclopedic memory for all the letters of the Torah also kept track of lineage which included marriage and divorce. Many commandments and their subsequent laws dealt with the relationship between man and women; though, when it came to the act of sexuality there was no law, there were laws concerning when and with whom one could engage. The Rabbis watched over the people so men could leave their homes for months at a time and trust the Rabbis to compute the term of pregnancy consistent with the man's absence.

Besides insuring fidelity, the Rabbis identified and segregated parts of the community as commanded within the Torah. People with venereal diseases and other issues that plagued the body had to seek out the Kohan to make a determination on whether and for how long a period the person should leave the community. The most severe type of excommunication was the **Momzer/Bastard** who was not allowed back into the community until three generations, the great grandchildren of the Momzer had passed and only then would be allowed entry back into the community.

The idea of a married woman giving birth to another man's child was commensurate with serving idols; man and woman were to be a paradigm of Heaven and Earth where man gives and woman receives in the same way that God gives and creation receives—to seek sustenance from another is like idol worship and idol worship was the most repugnant of all misdeeds. To bring a **Momzer/Bastard** into society was considered the most egregious action against the Creator, even more than the worship of idols, since the guilt of the parents lives in the child; the humanity of the human being protects the **Momzer/Bastard**, but the law of the Torah rejects him for generations.

The laws of marriage are not the same for both man and woman; for example, man is commanded to marry and have children but women are not commanded—they marry and have children purely out of love; man is obligated to care for his wife—providing her with house and sustenance (receiving back the work of her hands, but if the women prefers to have job and support herself then she keeps all her money); man is permitted more than one wife but is prohibited from any sexual contact with another man whereas woman can have men or women, but when she marries she can no longer have other men—to do so would be adultery and a child born from such a merger would be considered a **Momzer/Bastard.**

The Rabbis did not accept the word of the mother who said she had not been with another man, insisting instead she had been spoken to by the angels and thus became pregnant nor did they accept the word of the midwives who accompanied her. If so, said the Rabbis, according to the law of the Torah where the seed goes in the child must come out—therefore if the child is conceived from her ear—the whisper of the angel and born from the ear—then the Rabbis would have believed her. The mother had a conventional birth and foregoing the sign of returning virginity, the Rabbis declared the child a Momzer.

The Momzer did not have to physically leave the community but he was not allowed to marry except another Momzer; after three generations the grandchildren of the Momzer would be allowed once again to marry into the congregation. The stigma was enormous. The Rabbis were being very strict, perhaps too strict, but they watched the child and were quick to bring him into their ranks seeing his brilliance. When he matured he became included in those called the **Tanoiyim/Teachers** who transmitted the law.

The law was divided up into 600 sections; Jesus was particularly adroit at **Negoiyim/ Leprosy** and was later quoted when the law was written down a century later. Only those whose words included the entirety of the knowledge in the most minimal amount of words would be included into the writing down of the Oral Tradition. Jesus had a high and respected position as author of knowledge, but the stigma of his birth as interpreted by the Rabbis was always between him and his people.

Jesus had an inkling as to what was going to happen, because he knew what was later to be written down in the Talmud-Book of Law: when asked, from where in the Torah do we learn **Tichiut HaMatim/Resurrection**; they forego quoting Torah but instead answered—from the way of birth—just as the seed of man goes into the woman quietly and comes out nine months later with great noise; so, how much more so, if, when the body is placed into the ground with much noise of sorrow will the body rise again in great joy. Jesus knew that the sign of his birth would be seen in his death.

It was during the Herodian times that Jesus lived. Herod had been a slave under the Chusmoniem a family of Kohanim who had taken power having ousted Greece; the law of the Torah concerning a non-Jewish slave contends a slave becomes Jewish upon being freed—the Talmud-Book of Law says the freed slave is the cruelest of people. So it was with Herod, having lived in the house of power once freed knew how to make alliances with Rome to maneuver himself. So the slave could replace the master.

The Talmud-Book of Law tells the story of a man who bought a slave to work in his house. In those days slavery was commiserate with permanent employment; whereas people were taught to trust in God they nonetheless had the right to sell themselves for as much as seven years to a stable environment that would provide food and shelter for him and his family. This would give the slave time and opportunity to rebuild his finances and free himself from the slavery of his obligated job.

In the Talmudic story, the slave lends the master an amount of money which the master returns through the hand of his wife. The slave keeps the woman and the money claiming that ruffians took her, but in truth he kept the wife and abused her. The slave convinces the man to divorce his wife and the slave would give the man an additional loan to pay off the marriage contract—which his does. When the loan becomes due and the master can not pay he must sell himself to the slave who now becomes the master. In the end the slave who used to be the master is told to bring a cup of wine to his new master—the previous slave who is now married to his previous wife. When he sees them sitting together a tear falls into the glass of wine and a voice comes out from heaven and says: I will destroy this house. Herod was the whip to punish the people and the precursor of destruction.

Herod had unleashed a wave of violence against the Rabbis fearing their erudition and their secrecy, but later repented his ways and asked the remaining Rabbis what he could do to restore his credibility—they encouraged him to restore the Temple which had fallen into disrepair since the time of the Greeks. Refurbishing the Temple was a way to rebuild that which he had torn down through the killing of wisdom. The Talmud-Book of Law says he who did see the Temple of Herod did not see beauty—like a static ocean wave hovering in the desert, nestled between two hills.

Along with the violent opulence of King Herod, the intrigue of political life which reached even into the Temple fermented distrust and seeded doubt; stories abounded about how the honor of being Kohan Gadol was being auctioned off. The practice became so rampant that before the Kohan Gadol entered into the Holy of Holies, the empty room where the Ten Commandments once stood, they would tie a golden chain to his foot. It was known that if the Kohan Gadol had even one stray thought while standing in that place he would immediately die.

Corruption abounded throughout the Land of Yisroel, but at the heart of it all were the Tzadoki and their creed: there was no reward for spiritual work. They favored making a division between the spiritual and the physical thus becoming a nation like all other nations. They rejoiced with the Greeks and later plotted with Herod; they were responsible for political reality inculcating itself in all aspects of Jewish life. The only fortress left was the Temple and even that was being corrupted.

It was up to the Rabbis to take action, but they did not; they stood by and was non- committal seeing both sides of the situation. The Rabbis were extremely secretive, speaking in metaphor or relaying messages through the interchange of letters and numbers. The strata of their interactions were so complicated that they encouraged suspicion. And Rome ruled through the sword.

Jesus as a child was drawn into the ranks of the **Perushim/ Explainers** and was said to have spoken to Hillel at the age of three. Hillel had begun a school of thought that would persist until today; he was famous for having taught the essence of the Torah: love your friend as you love yourself—the rest of the Torah is explanation of this concept. Hillel's rival school was named Shami; they were known for their strictness and Hillel for his kindness. Jesus had taken a teacher of extreme strictness, intolerant of blemish. The Rabbis had put upon Jesus the greatest of spiritual burdens and he had chosen a teacher of the highest regulatory temperament which was the deciding factor—too much severity.

The story is told that Jesus was accompanying his teacher; they were at a tavern and his teacher had made a comment about the beauty of the girl serving the food and Jesus responded acknowledgement. His teacher became furious. How dare Jesus, a Momzer, have his eye on a Jewish girl? His teacher demanded his student leave. The Torah stipulates in the case of asking forgiveness that three times is sufficient; if forgiveness is not won in the third attempt to ask for forgiveness then the burden to ask is lifted.

Jesus returned for his third audience with his teacher, if his teacher refused this time Jesus had decided to go off on his own. As providence would have it when Jesus returned his teacher, who was prepared to forgive him was in the middle of saying the Shema Yisroel and made a sign to his student to stay and wait, but Jesus understood from the subtleties of the of the sign—go and don't come back.

It says in the Talmud—Book of Law that Jesus had five disciples; having rejected the Tzadoki and been rejected by the **Perushim/ Explainers**, the Rabbis, Jesus went off to the Essences to establish his own school. He had learned much from the Rabbis and knew the secret to the power they held which was embedded in the Name. When Jesus went out to the people to unleash his power, the Rabbis feared—least their power be known.

Jesus was trying to forge a new path: power to the people—a thing neither authority, Rome or the Rabbis, wanted. They called him **HaNotzri/The Cast-Out** since his status as Momzer precluded him from partaking in life's pleasures; the Notzer was one who refrained from high forms of pleasure because of his own internal weakness which would lead him to misdeed.

Jesus was rebelling from the severity of the Rabbis by releasing secret into the world because he knew that in his death would be the greatest secret of all. He taunted the Rabbis until they killed him—put him to death as a false prophet. Though strict, the Rabbis rarely put anyone to death since it was their charge to find a loophole in the law that would allow exemption— except in the case of false prophecy. There was no greater offense than speaking false in the Name of God.

It was a delicate game in which Jesus and the Rabbis were engaged; Jesus understood the meaning of his predicament and knew his prophecy would be spoken at his death—it was the sign of his mother's virginity at his birth. First being castigated by the declination of **Momzer/ Bastard** and second being excommunicated by his teacher led him to want his death sooner than later. Being a Rabbi he knew how difficult it was to get the 71 members of the Sanhedrin to agree—particularly in a capitol offense.

It is said that Jesus wrote down the pronunciation of the YHVH on Y**om HaKippurim/ Day off Atonement** when the Kohan Gadol emerged from the Holy of Holies to pronounce the Name and all the people bowed down; later he cut open his flesh and placed the paper within and was able to fly in the skies above Yerushaliam. Rabbi Ishmael the Kohan Gadol took in his hands the Tzitz, a gold plate that he wore on his head where was written the Name YHVH, and he too was able to fly. The Kohan Gadol rose above Jesus, urinated upon him and brought him down to the ground where he was taken and brought before the Great Sanhedrin of 71 elders.

Jesus was charged with being a false prophet by using the power and secret of the Torah to do miracles like: feeding the poor and healing the sick. Jesus retorted that indeed the Torah commands us to feed the poor and heal the sick. But, the question was the use of miracle as the Talmud-Book of Law contends: even God uses miracles only as a last resort and even then the miracle is of the lesser degree. For example, one of the Rabbis lost his young wife who left him with an infant baby; the Rabbi was poor and prayed to God for means by with to sustain his baby's life and the man grew breasts. From this the Rabbis understood that since God always does the lesser miracle that for this man, to grow breasts was easier than getting a job.

The Rabbis contended that Jesus was blatantly breaking the laws of nature for his own aggrandizement. A large following had grown up around the Momzer—people of ill-repute and other cast-offs of society found solace in his persona and stories abounded. But, Jesus grew weary of this world and decided to deliver his message and leave. Through his antics in the air Jesus had brought forth the Tzadoki always ready to make trouble for the Rabbis. Jesus' message was for the Tzadoki but the Rabbis were going to help send the message and release the prophecy.

It was a known thing that the least of the Rabbis could raise the dead and they themselves were able to elevate into the heavens communicating with angles and souls, but many of the Tzadoki were hidden even from the Rabbis. Since Tzadokim are defined by his intention, one might suspect because of a peculiar behavior or subtle reference in the words spoken, but to know what is in the heart is a capacity reserved only for the Creator. Jesus had been sent to deliver a message that would be spoken upon his death. The Rabbis agreed to strangle him for his words. And that is what they did.

The laws concerning capitol crimes are explicit and adamant—the one being executed must not experience pain. The Rabbis did not like to put to death and rather preferred that God send the Angel of Death at the appropriate time, but there were instances of necessity and false prophecy was one of them. Death as instantaneous as it appears is much more complex than what ostensibly happens: consciousness is no longer housed in the body which becomes inanimate.

With Jesus two things happens: first, angry curses were choked from him and caught by a mysterious man by the name of John who wrote them down into what is know as Revelation. Though these angry words are delivered in the form and style of prophecy it is well known that negative prophetic visions of the future need not come true. Second, Jesus was the first to undergo **Tichiut HaMatim/ Raising of the Dead** and it is why John refers to Jesus with the curious title of "The first born of the dead."

The message in Jesus' transformation was simple: not only is there reward for the soul, but also for body—a clear message to the Tzadokim to check the premise of their beliefs. It says, in death the biggest discomfort is embarrassment for not seeing what is true. Much of what the Rabbis argued about was perspective—perusing the truth through many different perspectives with the objective to discover as many as possible. Not every path leads somewhere; some paths are blind allies, when you get to the end you have to return back to where you came from— repentance. God sent Jesus with a message for the Jewish People to quell the suspicion and hatred that had been festering for centuries, but neither the Tzadoki nor the Rabbis received and accepted the message. Jesus died, then grew from the ground a new body, an eternal body that walked into the heavens as the prophet Eliohu had done during the time of the First Temple. The Tzadoki had no place for miracles and the Rabbis refused to retract their decision as expressed by the following Talmudic story.

Once the Rabbis were arguing with a **Y'Chid/Individual**; though the Individual had great merit to his argument still the Rabbis, who were the majority, refused to adopt the individual's claim. Finally in exasperation the Individual challenged the Rabbis saying, "If I am right let this tree uproot itself from the ground and walk away." And that is what happened, but the Rabbis were unimpressed.

"If I am right," maintained the Individual, "let the river run upstream." And the river did but the Rabbis remained unimpressed. Finally the Individual said, "If I am right let God be my witness." And a voice permeated the heavens and gave witness. The Rabbis lifted their faces to heaven and said to the Creator: stay out of this, we make the law. And so it was and God said: today my children have defeated me; then looking at the angels said: Just shut-up.

Jesus was that individual who uprooted himself from the Earth and like a river going upstream returned back to the heavens where God was witness to what he had done. A conglomeration of bad judgments and flagrant misunderstandings led to one of the oddest stories in the annals of the Torah whose only reference are subtly hinted at in Oral Tradition like a family secret that no one else would understand, but more explicitly recounted in the Koran.

The Talmud-Book of Law says Jesus had five disciples who conspired to take on the simile and manufacture a story which they sold to the Tzadoki who in turn, after a couple of centuries, sold it to Rome. The disciples who could not claim prophecy risking death like their master called their writings: an inspired work. This fabricated story became the cornerstone upon which the Jewish People would suffer for two thousand years.

HISTORY

Born around the time Jesus died, Josephus was a young general who when surrounded by the enemy decided it was better to commit suicide than to be taken hostage, but as he saw the men around him dying he began to change his mind and finally when the minority who wanted to live rather than die became the majority they gave up to the enemy. Josephus as he explains in his own words had a sudden illumination, an epiphany about the future of the world and how it was as the Torah said: as Yisroel falls Rome arises.

He saw it was better to become a part of Rome since it was obvious where God was pointing the Jewish People; by becoming part of Rome the goodness of the knowledge of the Jewish People would ultimately influence the world to seek God. What the Jewish People could not accomplish from their own land—to be a beacon to the world—could be achieved as individual ambassadors spread throughout the Earth.

In the Torah the Jewish People are known as **Anshie Orif/Stiff-Necked** and it was difficult convincing them; eventually Josephus becomes interloper for Rome accompanying Titus on the siege of Yerushaliam. Unable to convince the rebels to disarm, he witnesses the destruction of the Temple commenting on how the Kohanim jumped from the Temple wall into the fire, but Josephus never mentions a word about Jesus.

Later Josephus is commissioned by Rome to write a history of the Jewish Wars and later a history of Jewish people which he calls: Antiquities. Though he meticulously details the numerous accounts of intricate dealings within the Roman province of Judea, thoroughly elucidating political alliances and recounting intrigue but Josephus does not utter a word about Jesus. Though Josephus does not mention Jesus he does mention John the Baptist saying: he is madman.

Also Pliny the Elder noted Roman historian who witnessed the destruction of Pompey two years after the destruction of the Temple also says nothing about Jesus. Pliny of Egypt says nothing of Jesus. In all the annals of Roman rule nothing is mentioned of Jesus, no such crucifixion recorded and no such incident is filed in Pontius Pilate's court. In two thousand years not one piece of evidence has surfaced in corroboration of this manufactured story.

The internal episode which occurred within the Jewish People was not recorded by Josephus for though having studied with the **Perushim/Explainers**, the **Tzadokim/Deniers** and the Essenes (a non- violent sect of Jewish People)and certainly had knowledge related to Jesus sought not to disclose it—anyways miracles were considered dubious and did not take place in history. The most obvious reason why the story of Jesus' crucifixion stands without corroboration is simply because the story did not happen, but the man Jesus did exist—misunderstood and inappropriately slain by his own people.

His message was one of conciliation between the Jewish People to heal the wound of the exile caused by the infliction of doubt. Jesus was sent by God to show the Jewish People how to live the 613 **Mitzvot/Commandments** of Torah within the exile and through it achieve for the whole world eternal life—not just for the soul, but even for the body. It was a simple message meant to strengthen the Jewish People during the long exile but in the end, because we did not merit, it made the chasm even greater.

Rav Tzadok (no relation to Tzadoki) began to fast when he saw what had transpired. The Talmud says he fasted for forty years and then the Temple was destroyed because there was hatred between the Jewish People and the hatred had no reason. Hatred without a reason will lasts for a very long time. A remnant of this hatred is yet seen today in the adaptation of the Jewish calendar for those living outside of the Land of Yisroel where one day holidays are celebrated for two days according to the laws instituted by the Rabbis.

During the time of the Temple through means of flags and fires the news of the new month would be transmitted to the outlying areas as to coordinate the observance of holidays falling in that month. The Tzadoki took leave to signal the new moon, which predicates the day of the holiday, a day early thus breaking the coherence and unity between the prays of the Jewish People within the confines of Yisroel and those living among the nations. Instituting a two day holiday outside of the Land of Yisroel robbed the Tzadoki from indiscriminate acts of hatred.

ONKELOS

A relative of Titus who led the Roman army to destroy the Temple, Onkelos had been instructed by his family to go out into the world and find something which was of great value and dedicate himself to what he found. Onkelos found Torah. When the family heard what he had done they sent guards to bring him back, but Onkelos explained to them how wonderful and wise is the Torah and they stayed and became Jewish.

Again, the family sent forth soldiers to bring back Onkelos but this time they were instructed not to speak with him. As they were silently escorting out of his dwelling, Onkelos reached up to touch the mezuzah affixed to the doorpost. One of the soldiers asked what he was doing and Onkelos responded: you put your god in a house and worship them from without while the Jewish People sit within while God watches over us. When the second battalion did not return the family gave up hope.

Onkelos is remembered for hid dedication to the Written Torah translated into Aramaic called Targum Onkelos and until today these words are reproduced alongside the written text of the Torah. Before actually becoming Jewish, while he was still weighing the advantages of giving up the Seven Commandments of Noah for the 613 obligation of the Jewish People he decided to ask the advice of those who were dead. First he brought up the soul of his relative Titus who burned down the Temple, next he brought up Bilaam who had been hired to curse the Jewish People—he tried but failed. Lastly Onkelos brings up Jesus.

Both Titus and Bilaam agree that it is better to fight against the Jewish People because it will bring with it power, money and high position. Each is asked what is their punishment for the deeds of their lives? Titus announces that he is judged by his own words: before his death he ordered his servants to burn his body and spread them upon the sea so the God of the Jews would not get him—every day he is gathered and burn anew. Bilaam, who introduced harlotry into the Jewish People, is boiled in semen.

Jesus encourages Onkelos to follow in the Jewish path whose ways are difficult but whose ultimate reward is beyond all others. Asked what his punishment is for the deeds of his life, he reveals that he seethes is excrement; the Rabbis say it is because he ridiculed the strictness of the Rabbis but others say because everyone puts their shit on Jesus and he just seethes.

In the end, and the Rabbis agreed as witnessed in the Talmud-Book of Law, the Temple was destroyed because the Rabbis were too strict—so strict that when action was needed to be taken they could not muster the strength to act. Coming at the end of the 1000 years of **Netzach/ Victory** was the defeat—a result of the deficiencies of the Jewish People. If perfect, the perfect thing would have happened.

Instead, something else happened—something that would help fix generation after generation so Moshiach can come.

Chapter Eight

Retreat

KOMTZA

The Talmud-Book of Law is divided into two aspects: law and story. It is said that if you want to know what to do learn the law, but if you want to know what is, learn the story. The stories related in the Talmud are fictionalized accounts of what happened. Though truth is certainly stranger than fiction it is also true that fiction can be truer than truth to make up for the endemic flaws in the human condition.

The story of Komtza and Bar Komtza is an engaging saga which puts the blame for the destruction of the Temple squarely on the strict shoulders of the Rabbis and their predilection for not taking action. Talmudic stories are written on four levels like Torah; there is the story, the law derived from the story, the extrapolated idea which comes forth and the secret of what really happened.

In the story of Komtza and Bar Komtza a man gives a party and directs his servant to go bring his friend Komtza, but the servant mistakenly brought his enemy Bar Komtza. The host insists that his enemy leave his party. Not wanting to be embarrassed Bar Komtza offers to pay for what he eats, but the host demands he leave; he offers to pay for half the party as not to be disgraced and again his offer is turned away. "I will pay for the entire party," says Bar Komtza and the host picks him up and throws him out while the Rabbis watch.

Since the Rabbis did not restrain the host from causing him embarrassment he would go to the governor and inform on the Jewish People. The governor is reticent and demands some corroborative evidence to substantiate his claim. "Give me a cow to send to the Temple for sacrifice and see what they do," challenges Bar Komtza—he knew the subtle difference between what Rome considered acceptable and what the Rabbis would tolerate.

Some say he made a spot in the eye while other contended he cut the lip. The Rabbis inspected the animal sent by the governor and determined it was **Posel/Unfit**. The danger of refusal was evident and the Rabbis considered sacrificing the animal anyways for the sake of peace with Rome but other Rabbis voiced the concern that to compromise the law was strictly forbidden since people will say the Rabbis compromised the law. Some thought to kill the man who brought the animal, but others responded the people will say he who brings an unfit animal will be killed and so the Rabbis did the law in its strictness—that which was unfit was not offered; because of this, the Talmud concludes, the Temple was destroyed.

The story never explains why in the beginning it claims the Temple was destroyed by Komtza and Bar Komtza. It is clearly understood that Bar Komtza played an important part of the story, but what did Komtza do to exact such blame? It is explained that he did not come to the party and waited to be invited and the Midrash- Book of Metaphor says the Temple was destroyed because of indiscriminant animosity practiced among the Jewish People called **Sinot Chinom/Free Hatred**.

OCCUPATION

Five hundred years had elapsed since Babylon had destroyed the First Temple and taken the people from the land. The first five hundred year period of the fourth millennium called **Netzach/ Victory** had the Jewish People living in prosperity on their land in relative peace producing great works of prophecy and the foundational royal dynasty that would extend to Moshiach. The second 500 year period was a process of disintegration; though the Second Temple was built and rebuilt beautiful and gracious, yet it lacked the soul personified by the sapphire cube of the Ten Commandments.

Though the people were restored back to the land it was with occupation: first Persia, then Greece and at last Rome. The Persians were elegant people led by leaders pursuing truth; the Greeks put beauty before truth; and the Romans cared for neither—conquest was its own reward. Life under Rome was brutal—the ancient curse of Esau was becoming evident. The Jewish People had become belligerent and Rome was not to be trifled with.

Vespasian was sent to quell the rabble as the Talmud-Book of Law relates. An insurrection was growing within the walls of Yerushaliam that threatened to destabilize the region. At the news of Vespasian's intended venture, the Rabbis encouraged the people to make a siege out of it hoping after a few decades the enemy would go away. The leaders of the city pledged enough wood, flour and oil to sustain the city for 21 years, but the rebels who wanted to fight burned it all down and famine began.

Rabbi Yochanon ben Zachi head of the Sanhedrin feigned weakness, claimed death and was smuggled outside the city walls where he sought out Vespasian. Upon meeting the Rabbi heralded Vespasian as king because Yerushaliam can only fall to a king. At that moment a currier entered with a message: the king of Rome had died and the Senate had voted Vespasian to be Rome's next king.

When Vespasian received the news he was in the middle of putting on his shoes; after the news he was no longer able to put on the second shoe or take off the first. Rabbi Yochanon explained that good news makes the marrow thick and suggested he bring in a soldier with whom he was displeased to shrink back the marrow. Seeing the great wisdom of the Rabbi he asked him if you knew I was king and you were so wise, why did you not come before? Now I must break the barrel of honey to kill the serpent.

The Rabbi had no answer, but instead pleaded to be given the city of Yavna where the Rabbis could continue their work writing Mishna which is what he did; Vespasian did not attack Yerushaliam, but he promised to send his son who would. Particularly interesting about this story is its repetition by Josephus who explains his own meteoric ascent into Roman influential positions by means of the same story—claiming to have escaped and come to Vespasian to prophesies his impending appointment as king to Rome.

JOSEPHUS

When the First Temple was destroyed by Babylon the eyes of the king were put out and he became blind; when the Second Temple was destroyed by Rome, the light of the Temple was distinguished and the world fell into darkness—with the obliteration of the light one historical eyewitness who wrote in length about the legacy of the Jewish People, was left: his name was Josephus.

Of the many historians that populated the Middle East, only the works of Josephus remained in its completion because he wrote under the auspices of Vespasian and Titus: emperors of Rome from the same family. It is said: the victor writes the history meaning the victor selects which account of the history will be preserved and which will be eradicated. What was it about Josephus which won him this acclaim?

Josephus has an articulated creed, a blend of Judea-Roman influences woven together; he sought to speak for the Jewish People, including in his history a retelling of the Torah. Josephus, who had learned with both the Pharisees and the Tzadoki, went off to live with the Essenes in much the same way Jesus was said to have gone off and lived with the Essenes. He retells the story of Rabbi Yochanon ben Zachi who prophesied Vespasian's ascent to king of Rome and thereby won freedom for the Rabbis, but places his own personage in the Rabbi's place.

Josephus had great pride in being Jewish, devoting a large section to the retelling of Masada, though insignificant in the overall campaign to subdue Yisroel, speaking to the proud spirit of the

Jewish People who preferred to die rather than to be taken captive by the Roman army—the night before their inevitable capture the group committed suicide after two years of battle.

Josephus, who traveled with Vespasian and accompanied Titus at the sacking of the Temple, described how the Kohanim jumped into the fire and was witness that the destruction occurs on precisely the same day, the ninth of Av, when 500 years previous the First Temple had been destroyed by Babylon. He too was a Kohan from the princely tribe of the holy people; it was for this reason that he was sent from Yerushaliam to the Galilee in the north to try and make peace with the Jewish rebels who wanted to fight with Rome.

The Kohanim and the Rabbis understood it was better to live with Rome, lie down with the lion, and wait for them to fall as Greek had fallen. In the middle of the battle Josephus has an epiphany and sees Yisroel being swallowed by the lion, ingested into Rome. Fathered by Esau, Rome has no spiritual heritage having sold that to his twin brother Yaakov—progenitor of the Twelve Tribes of the Jewish People. The Royal family of Rome chose Josephus' history and his conspicuous watering down of Torah, his dissolution of miracles and eradication of the spiritual influences that brought about historic events as a conciliatory position somewhere between the Pharisees and the Tzadoki.

Josephus, in many ways, is the link between what was and what is. From this point, because of the destruction of the Second Temple and the dispersion of the Jewish People Rome had decreed life without God, instead they turned to their father Esau and his connection with the planet Mars. Rome chose Josephus to disengage God from the world by minimizing miracles and limiting life to our little earth and the seven celestial objects that are readily seen surrounding the earth: sun, moon, Mercury, Venus, Mars, Jupiter, and Saturn. Roman life precluded everything but what could be visibly perceived or calculated.

ZEALOT

Besides the three groups delineated by Josephus: Pharisees, Tzadoki and Essenes, there were the Zealots—a rebellious group who saw in the Torah the responsibility and obligation to rid the land of the invaders. The zealots were very popular with the people because they reminded them of their illustrious past and how God had told the Jewish People they would be the head of nations. The land of Yisroel was sacred—it was the land given to Avraham from God and passed on through the Twelve Tribes.

Only a remnant of the Twelve Tribes was left: Levi and Yehuda plus the land were in turmoil encroached upon by the Roman Empire. Jewish ways were being eradicated while Roman gods were on the threshold of the Temple waiting to be installed. The people were furious and the rhetoric of the Zealots was like a match on dry hay. There was less than 200 years left until the end of the thousand years denoted as **Netzach/Victory**.

It was a time of great confusion and multiple divisions with variant and opposite wrath incurring perceptions inspiring hatred without reason. The over thousand year history of Yisroel was coming to an end—defeat at the end of the thousand years of Victory was inevitable. The Zealots could not accept the fact the Jewish nation was coming to an end. How could this be?

Where was Moshiach? Some said Moshiach came forty years before but the Rabbis killed him.

Yisroel was coming apart at the seams; it was the love of one another that stitched the people together, but the boiling hatred in the cauldron of oppression was tearing the people apart. The Zealots, likes all zealots, could not accept reality and through the force of their will, the strident belief in their convictions and their willingness to die for their cause made them to be a formidable force.

The Tzadokim meanwhile were making alliances with Rome and were happy to become a secular state while the Rabbis, who were the real seat of power reminded everyone that as Dovid had said: even as the sword is upon you neck, pray. Some prayed to an extreme like the Essenes notorious for their cleanliness of body and spirit through strict obedience and a vegetarian life style.

The words of the prophets were being interpreted by each group who saw their vision being prophesied. Rav Tzadok who had been fasting for forty years in hope the Temple would not be destroyed was so thin that at night when he broke the fast to suck the honey from a date it could be seen the juices transfusing through his veins. Lastly a bright comet with a vast tail hung in the sky above the Temple as Titus approached with his army to destroy the Temple.

The Rabbis had encouraged the people to wait out the siege, but the Zealots burned the stores and a famine ensued. On the 17th day of the summer month Tamuz Titus broke through the walls that surrounded the city and was steadily making his way towards the Temple. A comely woman who was hungry, never before faced by famine, found the date from which Rav Tzadok had sucked the juice, she tried to eat it and died—fulfilling the prophesy regarding the destruction of the Temple: even the comely amongst you will be ravaged.

TITUS

When the Roman general Titus left Alexandria to quell the rebels of Judea, he took with him as translator and conciliator Josephus friend of Vespasian king of Rome. Josephus was thoroughly dismissed by his countrymen for his complicity with the enemy and his message of reconciliation certainly fell on deaf ears. The pride of the Jewish People was being tested; once again God had sent an army to destroy the house built by God's command and the Jewish People were to be orphaned into the world.

The seventeenth day in the summer month of Tamuz struck a particular discordant note since it was on that day Moshe descended from the mountain and found the people had built a Golden Calf which they installed and worshiped in place of God; as a result he threw the Ten Commandment engraved by the finger of God to the ground and they shattered. It is said in the Midrash-Book of Metaphor that if the first set of Ten Commandments had not be shattered then Moshiach would have come and none of this would have happened. It was an ominous sign that Titus had broken through the wall on that very day.

The First Temple had been destroyed on the ninth of Av which was only three weeks hence. Bickering between various factions broke out and bitter words were exchanged; when the warriors of Rome breached the walls of the city and found the people engaged in argument and accusation they killed without mercy. Then Titus turned his eyes on the Temple and gave the order: it should be burnt.

The Kohanim had gathered together and decided to use their collective abilities to ward off any incursion into the Temple by Rome—they would concentrate on a particular psalm relevant to the situation—with the psalms completion angels would be drawn down to protect the Temple, but three words before the end the soldiers breached the inner wall and broke their concentration. Seeing the flames lick up as the limestone ignited the Kohanim leapt into the flames.

Titus exalted by his victory claimed his place in history by bringing a prostitute with him into the burning Temple and as the flames of victory dance around him he tore off the curtain that separated the Holy of Holies from the rest of the Temple and threw it on the floor of the room where the spirit and matter merged. Upon the holy curtain he unraveled the scroll of the Torah and then he stripped naked the woman and had sex with her there in that place of holiness.

The Midrash-Book of Metaphor says, before tearing down the curtain Titus first stuck his sword through the curtain and the curtain bled—from this Titus thought that he killed God. He packed up the golden vessels and the Golden Candelabra and brought them back to Rome where to this day he is pictured on what is known as the Arch of Titus a representation of the seven branch candelabra being brought into Rome with the Jewish People in chains, their heads bent in submission.

With the flames of the Temple the light of the world had been put out and the golden vessel that housed the light was buried in Rome.

TUBA AV

The Jewish calendar is calculated by the movements of the moon; each month begins with a celebration concurrent with the initial phase of the month; R**osh HaShana/New Year** is celebrated on the first of the autumn month called Tishre when the moon is completely obliterated by the shadow of the earth. The full moon for the Jewish People was during the time of **Shlomo/Complete** who built the First Temple where the pillar of smoke never altered and fire in the form of a lion descended from the heavens to devour the animal offering had waned.

There is a gentle wisdom found in the phases of the moon that teach us how to rejoice in fullness and to let go, disappearing into blackness only to reemerge. Once the destruction of the Temple was complete and the people saw that the moon had waned and light was quickly in decline they became resigned to their fate—the encompassing darkness of exile. The Temple was no more, but now that it was gone, certain things were clear.

The Second Temple had been destroyed on precisely the same day as the First Temple as a sign that providence is more than fate—it is a message. The Talmud-Book of Law says the greatest Jewish holiday of the entire year is **Tuba Av/Middle of Av** (later summer month)for as bad a day as was the 9th ofAv the 15 of Av is as good. The men of the **Midbar/Wilderness** who wandered with Moshe for forty years destined to die by the age of sixty would lie down in a grave the night of Tuba Av and if they awoke in the morning it meant they would have another year of life.

Later Tuba Av took on another life affirming event; single women would trade clothes with one another, each one dressing down for the benefit of her sister so in the end the coalition of women who marched through the streets dressed in white with their timbre and their drum daring the young men to enter into matrimony.

The Talmud-Book of Law recounts how the women of pedigree would say to the boys: turn your head to lineage and think of your future children; the women of beauty would say to the boys: turn your head to looks. The women with neither family lineage nor looks said to the boys: do it for God because you were commanded. The cabalists say the reason why this holiday is greater than all the other holidays is because it comes directly after the destruction of both Temples, on the fullness of the moon to remind the Jewish People: this too, the dark exile, will pass.

The Temple was destroyed and with it the capacity to perform many of the commandments; the Sanhedrin was abolished and the land of Judea became like any other protectorate under the Roman Empire. Av, the month signified by the lion was chosen both for destruction and redemption. The moon could not face off against the sun so she made herself smaller; as had been the blessing of Yaakov: as Yisroel goes down, Esau ascends.

FLAVIUS

The Roman family responsible for the sacking of the Temple, Vespasian and his son Titus were not from the royal Roman blood that led back to Esau, but rather conquers who through the force of their courage in battle had risen in the Senate until they captured the crown. With Josephus as their spokesman they established a creed based on world conquest. Josephus had warned the Jewish People that no power was equal to Rome and no nation had ever defeated Rome.

The Midrash—Book of Metaphor says: if the seas were ink, and if the reeds in the sea were pens, and if the heavenly expanse was the paper there would not be enough to write the workings of Rome. Rome excelled in the manufacturing of technology, with an innate ability derived from the planet Mars to accrue and put things together; the brutal discipline required making the apparatus work, Rome forged out of blood and steel producing the invincible Roman Empire.

The administration of the army was done through the government whose economy—the coin of Rome—paid for the many advances like roads and aqueducts, but it was war and conquest which drove the economy. More than the industry of Rome was the religion of Rome. Esau was the Patriarch from which they drew their blood and Esau had been blessed to live by the sword. Esau had little use for the spiritual having sold his birth right to his brother.

Lacking an indigenous connection with the spirit world led Rome to adopt the ways of the people they conquered, for example the Greek gods were shaved and renamed by Rome; Ares the Greek god of war was cruel yet ironically praised for his great beauty; his throne on Mount Olympus was covered with human skin. In Rome Ares was called Mars; the father of Romulus and Remus the mythological founders of Rome—twins from Mars like Esau and Yaakov. Mars even sports two twin oblong moons, one destined to crash into the planet Mars.

In the Zohar-Book of Secret states when Shlomo at the beginning of the thousand years **Netzach/Victory** married the Queen of Sheba from Egypt the ange**l Gabriel/Servant of Severity** placed a reed in the sand and that was the beginning of Rome from the blood of Esau and the native Etruscans, the Roman empire emerged a thousand years later to rule the earth. It was from the more agrarian Etruscans, that the Flavians traced they're lineage.

In order to tap into the power of Esau the Flavians were compelled to face the gods and make their offering. The ancient position of Pontiff had ceased with Julius Caesar; it had been his job as Pontiff to establish ceremony. By becoming emperor Julius Caesar had melded the two authorities into one and after his assassination the post of Pontiff was discontinued.

Vespasian and his son Titus understood the benefit of having a new history of the world that deemphasize gods and miracles—Josephus was just the right person with his diluted version of the Torah that would put Roman theocracy on the defensive. Domitian son of Titus and virulent hater of the Jewish People continued to support Josephus and promulgate his work in opposition to all other historians. Precursor to the Bible.

MARS

Romulus and Remus were the offspring from the gods Mars and Venus; they were left to die on the palatine hill by the Tiber River according to mythological remains; fed by a wolf and eventually cared for by shepherds, the two boys decided to build a city. Each took a hill on the opposite side of the river, but then Romulus killed his brother Remus and on his bones **Rome/ High** was built during the time of the First Temple.

This story of one brother killing another is physically portrayed in the bizarre anomaly presented by the two moons of Mars: **Phoebus/Fear** and **Demias/Panic;** each of the two oblong moons is only approximately ten miles across, but the larger of the two, Phoebus, is falling precipitately close to the surface of Mars and will shortly, is astronomical terms, come crashing down. Because this story is inculcated into the very marrow of Rome, the god of Mars plays a particular significant role in the workings of Roman Empire.

The soul of Esau not only came from Mars through his grandparent Avraham and Sara but also exhibited attributes associated with the peculiarities and uniqueness of our sister planet. He was red like the planet Mars and he was already made, thus his name **Esau/ Already Made** at birth with teeth, body hair and a beard; Mars is an internal world where water is inside the rock and growth is a matter of accretion. Mars is also the planet where the Tower of Babel was built and destroyed breaking language and giving Avraham the power to come to the Earth and introduce new advanced languages that would allow for easy conquest of opposing people without warfare.

Esau who lives by the sword combined with the Martian attribute of **Gevorah/Severity** and Etruscans sensitivity to art blended together at the foot of the Mediterranean was the flesh and bones of Rome devoted to Mars. The word Pontus as in Pontiff means a bridge spanning a river that separated the earth from Mars—a place where souls could escape reincarnation and be reborn with great advantage. It was the Pontiff who conducted business between these two worlds.

The Zohar-Book of Secret identifies Mars as the previous world known as Tohu from where Earth was seeded with life. The animal in particular exhibits traits and strengths derived from the world of Tohu and as a result always mates from the back; in the world of Tohu accretion happened one behind the other. The animal nature in the human being derives from the virtues of Tohu and when combined with the virtues of Earth, the external world as represented by the human being who mates face-to-face, fixing occurs.

Rome began brother murdering brother and became an empire through the rigidity of discipline and the muse of philosophy learned from the Greeks; Rome became the cosmopolitan nation beneath the red glow of the planet Mars. Avraham's son Yitzchak wanted to bless Esau before Yaakov because he was the first born, as indeed **Tohu/ Chaos** preceded this world called **Tikun/Fixing**, and wanted to fix Tohu in its essence by blessing his eldest son; his wife, Rivka, tricked her husband into blessing Yaakov the younger son because she understood that Esau would have to be fixed by his opposite—his twin brother.

At the time of the destruction of the Second Temple there were two Jewish influences operating in Rome: the Rabbis and the Tzadoki. The Rabbis were concerned with maintaining coherence and stability so the arduous task of writing down the Oral Tradition could be accomplished before the Jewish People were removed from the land. The Tzadoki, on the other hand, were making political alliances with Rome and secret agreements that saw the transfer of spiritual and occult information concerning the power of the planets, in particular the planet Mars.

The Jewish People were presenting two faces to the world: one was the severity of the Rabbis and the meticulous nature of the law while the other, the Tzadoki, spoke in whispers stemming from vague intentions laced with ambiguity.

THE RIFT

It is established early on in the Talmud-Book of Law to be stricter than the Rabbis is too strict. Strictness is a function of contraction and it is through contraction that the essence is pressed out. The Rabbis squeezed every letter of the Torah to try and bring out its essence. The Rabbis established a border beyond which more strictness would break the Torah. The way the rest of the Jewish People lived the Torah established the other boundary of how far away one could be and still be within the realm of the Torah. The Tzadokim pushed the Rabbis to be stricter than the Torah provided and at the same time encouraged the people to be more lenient.

There are many examples throughout the Talmud-Book of Law of the kind of discussions that went on between these two sides: the devious questions posed by the Tzadoki and clever answers of the Rabbis. But, perhaps nothing is more indicative of the rift between the two than their stance over idol worship. Both agree, bowing down and prostrating one's self on the ground is idolatry, but the Tzadoki allow one to nod their head to the idol.

The law against idolatry is the stalwart of Jewish prohibitions; the Jewish People are commanded to die rather than bow down to a thing or object representing the Creator of the Heavens and the Earth. A nod to idolatry was consummate to the worst Jewish crime, but it was precisely this nod which bridged the gap between these two nations. The Rabbis' relationship with Rome was a tacit agreement of ever-changing boundaries that the Jewish People were expected to adapt to without complaint while the Tzadoki were ingratiated into

Rome through the inner power of offerings and the hidden knowledge deduced from ancient insights.

Both the Tzadoki and the Rabbis had advanced conciliation and opposed war, but once the war happened and the destruction occurred the hatred and distrust between the two camps only added to the woes of the nation. The Rabbis had taught to serve God as if there was no reward; to be a servant, like an animal takes the yoke—not to care about pain or even think about servitude. The Rabbis showed how to dissolve into the light through the power of the Torah and the 613 **Mitzvot/Commandments**.

The Tzadoki in the subtlest of ways struck at the very foundation of Jewish existence and dared to make a change in the Torah. It was a chaotic time and out of the turmoil and the expansion of Rome a Torah was introduced into the world that differed by one letter—the letter Hey was changed to an Aleph in the word **Daka/Crushed** speaking of one whose testicles are crushed—the source of strength in the human male.

The Tzadoki destroyed the Jewish People from within but the Rebels destroyed the Jewish People from without leaving the people naked against world—thrown out from the land to wander in the darkness of the exile.

Chapter Nine

Sexuality

REBELS

Rome had struck at the very heart of the Jewish People by burning the Temple to the ground, but the limbs still moved and the body of the Jewish People vexed with pain sought relief in the words of the rebels who espoused revolution against Rome—though no nation had ever defeated Rome. But Yisroel had a secret weapon named **Bar Chochba/Star Son** who was an amazing warrior, plus there were things about his life—when he was born and different incidences that had happened that convinced the Rabbis that indeed this was Moshiach come to deliver the people after the destruction.

Rabbi Akiva, the greatest of the Rabbis, had even put a crown on his head, but later had to recant and admit he was wrong. The Talmud-Book of Metaphor tells the story of how Bar Chochba/Star Son prayed to God at the brink of the war that pitted himself against the Roman army and said: I do not ask to be helped only not to fight against me; at the same moment the Roman general prayed to God and pleaded he should not be remembered as the general who commanded an army but lost to one man. The Roman prayer was more pleasing before God than the Jewish prayer and as a result when Bar Chochba went to relieve himself a snake frightened him and he lost control of his bowels which spilled out onto the ground and he died.

With the defeat of Bar Chochba came the end of an era; for over 1000 years Yisroel had lived on the land and observed the 613 Commandments and now that time was over. The year of the defeat was 3880—there was yet another 120 until the year 4000. It was known that the first two thousand year period represented the infusion of life from Tohu which ended when Avraham left Ore Chasdim and came to the earth beginning the two thousand years dedicated to the Torah which ended 120 after the destruction of Yisroel at the defeat of Bar Chochba.

The last two thousand year period, according to the Talmud-Book of Law, is of Moshiach which Rome adopts as his banner. Interestingly, Rome counts 2000 years on the Jewish calendar in the year 5760 precisely 240 years until the year 6000. It is the law that if you steal you must pay back double—the destruction of Yisroel by Rome happened 120 years prior to the year 4000, as a result Rome's rule ended 240 years before its appointed time. The equality with which the Creator governs the creation is exact.

The year six thousand marks the beginning of the Thousand Years of Peace known also as the thousand Years of Woman which will have been brought about through the coming of Moshiach. Moshiach will bring light and the nations will open their eyes and the holiness of the Holy of Holies will extend to the whole Temple whose holiness will extend to the whole city whose holiness will extend throughout the country and all the land of the world shall become as the land of Yisroel—observing the commandments to let the Earth rest every seven years.

ELIOHU

In the very beginning of the Talmud-Book of Law relates a story about R. Yosi needing to pray the afternoon prayer enters into a structure recently destroyed by the Romans. At the end of his prayers, at the entrance of the hovel, he meets Eliohu the Prophet five hundred years earlier instead of dying and being buried in the ground like the rest of the human beings died and rose up into the heavens.

The Midrash-Book of Metaphor says because Eliohu remained in the womb for twelve months instead of the normal nine months his body was particularly refined and because he led an exemplary life at the end he was allowed to enter heaven with his body. Since he remained in his body he could return to Earth and blend in with the populous. According to the Talmud-Book of Law, Eliohu frequently met with particular Rabbis and even had appointed times study with them the Torah.

Because R. Yosi was the head of his generation, and like everyone of that time expected Moshiach eminently, he felt the necessity approach God with his expectation; as a result R. Yosi was met at the entrance of the hovel to deliver a message in the form of a teaching—he explained what happened in his own words: Once I was going in the way and I entered into a hovel, one of the hovels of Yerushaliam, to pray; came Eliohu and watched me from the entrance and waited for me until I finished my prayer. After I finished my prayer he said to me Shalom Alechim Rebbe and I said Shalom Alechim Rebbe and Mori/My Teacher. He said to me: why did you enter into this hovel? I answered him: to pray. He asked me: why I did not pray on the road? I answered him: I feared perhaps the people passing by would interrupt.

Eliohu told me to pray with brevity and in that time he taught me three things—he taught me not to enter into a hovel, to pray on the street and to let my prays be short. Then he asked me: what voice did you hear within the hovel? I answered: I heard a **Baat Kol/Spirit Voice** that murmured like a dove and said—Oiy! I have destroyed my house and burned my vessels and sent my children into exile with those who worship stars.

R. Yosi knew it was exile until Moshiach came and wanted to remind God by praying in a hovel. Eliohu came to teach R. Yosi the danger of entering into that which is destroyed and that there was great virtue in praying on the road, but keep it brief—the time for Moshiach was not yet.

THE VOICE

Once there was a laundryman who lived in the time of the great Rabbis like R. Yosi; the laundryman, in the Talmud, is always referred to as a person of little intellect and generally uncouth since their worked required them to deal in dog shit to extract the chemical needed to temper the mixture of lye and fat in order to the soften the launder in those times. It was in the morning when the laundryman heard the **Baat Kol/Spirit Voice** that spoke and said: whoever goes to the funeral of Rebbe will inherit **Olam HaBa/Next World.**

Spiritual voices were frequent in those days and the laundryman happily decided to go but needed to first finish his work and as the day progressed more and more laundry accumulating until he looked up and saw the sun reclining and knew he had missed **Olam HaBa/ Next World**. Distraught, he ascended to the roof of his house, threw himself over and died. Another **Baat Kol/Spirit Voice** pieced the air and said: he also gets **Olam HaBa/Next World**.

It was hard to know in the chaos of those days what was right and what was wrong. The people of Masada had suicide rather than be taken captives and made into slaves. The Rabbis had retired to Yavna to write down the Oral Tradition, the Kohanim and the whole tribe of Levi from which Moshe and Aaron came was decimated, the people were in upheaval-the years of **Netzach/Victory** were over; the sun of the Jewish Empire was setting—darkness was settling upon the Earth. The Shechina whispered in the wind and said: I will go with you where ever you go.

It took a few centuries for the Jewish People to assimilate into the world and those who remained on the land held on to the last glimmer of hope which would exist forever within the heart of the Jewish People—Moshiach. Throughout the Torah, Moshiach is never mentioned—it is the hidden secret that everyone knew, but no one knew what it is. Moshiach is the expression of God's essential Paeleasure at the culmination of creation, a gift from God. The Jewish People pray for Moshiach not to Moshiach.

Moshiach was better than this world and better than the next world. Moshiach could have come at the end of the Sixth Day of Creation if Adom had not eaten from the **Atz Chaim/Tree of Knowledge Good and Bad** and the world has been waiting ever since. Moshe would have been Moshiach had not he hit the rock and he hit the rock because the people murmured against him. Jewish history littered with 0people who through their deeds and their essential nature could have been Moshiach given the right circumstances.

There were some amongst the people who murmured and said: Moshiach had come and the Rabbis killed him.

THE VOID

In the four levels of the Torah: **1**) Written Torah, **2)** Talmud-Book of Law**, 3)** Midrash-Book of Metaphor and the **4)** Zohar-Book of Secret there is very little written about Moshiach himself other than he is male, from the tribe of Yehuda—the lineage of Dovid. Some say he will be born on the day the Temple was destroyed—the ninth of Av. What is written about Moshiach are the circumstances surrounding his appearance; there will be the war of Gog and Mgog, but it also says that murder will be eradicated; state of the world will be chaotic "trembling" and they will seek a fish from the sea, but in vain—they will not find it. There are many such citing throughout the Oral Tradition which illuminates the circumstances surrounding the appearance of Moshiach and for millennium the Jewish People have interpreted world events in the shadow of these signs, but with all the calculations and aspirations, Moshiach still has not come.

The importance of Moshiach is he delineates the end of the sixth day and the beginning of the seventh—when Moshiach comes the workweek is over. Time will have a new perspective, God established life on earth to be in seven thousand year cycles—six of those cycles would constitute days of work to fix the severity of the previous world which had fallen into this physical plane and required changing by means of kindness: the worlds of **Tohu/Chaos** and **Tikun/ Fixing**. If the year six thousand enters in and Moshiach has not yet come then it will be a The Jewish People would have failed and God would be forced to send Moshiach afterwards, as promised, to establish peace on the Earth for a thousand years—**Shabbat/Day of Rest**.

After The Thousand Years of Woman and Peace the life force of planet earth will be transferred to another planet—probably Jupiter, as it says in the Midrash—Book of Metaphor: we will cast away the Earth as if it is an old shirt. In the law of the Torah even a slave is free from a work that is done in perpetually—everything must have a limit. Great people have lived during generations so worthy that the rainbow, meant to remind God not to destroy the Earth, did not appear in the sky—but still Moshiach did not come.

The cabalists maintain that Moshiach will not come until the world is rectified and the 288 sparks that fell to the earth are returned to Mars. It is only a few hundred years until the year 6000; Mars has just come closer to the Earth than ever before in 2003 and now is receding opening a void within the human heart longing to be filled with peace and graciousness.

It says in the Talmud-Book of Law: when the world trembles look for Moshiach because in every generation Moshiach exists and the trembling are the pangs of birth.

INCARNATIONS

It is said that Moshiach lives in every generation, but Moshiach himself does not know he is Moshiach. It is Eliohu, the knower of many secrets from the heavens, who comes first and identifies Moshiach. It seems Moshiach is very humble—not one of the famous people; it is with great trepidation that he engages the world as expressed by the following Talmudic story.

Rebbe was so accustomed to his meetings with Eliohu that when the prophet did not show on an appointed day and came instead a day later, Rebbe reproved the prophet and demanded an explanation. Eliohu explained that because he had to come from the heavens to rescue Yisroel Saba, A Jewish Elder, that he was unable to make their meeting, but since he had not kept his appointment he would give over a secret from the heavens to Rebbe.

Eliohu explained to Rebbe how to find Moshiach—just go to the walls of Rome and there, outside the gates of Rome with the other beggars you will find Moshiach. Rebbe asked how he will know one beggar from the next and Eliohu explains that all the beggars remove bandages from the sores inflicted by the difficulty of their lives, bathes the sores then dresses them, but Moshiach dresses one sore at a time in the event he will be called upon to come he should not be in the middle of a caring for himself.

Rebbe finds such a man in the outskirts of Rome; with the beggars he dresses his wounds one at time. Rebbe asked him, "When are you going to come?" "This day," the beggar answers; Rebbe quickly returns to go home and wait, but Moshiach did not come. The next time Rebbe meets with Eliohu he asks him what Moshiach meant and Eliohu replies: not today, but on the day that is appointed.

It does not matter who Moshiach is, it only matter that Moshiach wakes up and comes to the world. It is known from the prophecies and elucidations that just before Moshiach is revealed the world will experience birth pains which will be expressed in the arrogance of the leaders, the strife between countries, famine and pestilence—many hungry with no one to pity them.

Acrimony will grow within the family and the family will break apart. All sense of propriety will be lost. To balance the slate of this difficult time the learning of Torah will become rampant and many secrets will be expressed into the world. Miracles will abound.

The coming of Moshiach is always rendered in terms of birth and there is no birth without conception and no conception without sex. The cabalists inform us that every part of the metaphor has meaning—sexuality in particular has enormous ramifications because it defies the laws of logic. The experience of sexuality is akin to the coming of Moshiach. Both culminate in one moment in an unprecedented display from the source of pleasure, preceded by considerable work whose conclusion is ultimate letting go followed by relaxation and something good to smoke.

SEX

Sexuality is connected to the two sources of the soul: Will and Pleasure—prototype for male and female; because sexuality is connected to the essence it is through sexuality that essence is transferred from one generation to another. Sexuality is not logical— the nonconformist limb of the body; all other parts of the human form work in conjunction with one another, but sexuality overwhelms the normal functioning of the body focusing all energy and awareness between the legs.

The cabalists equate the lower part of the body to the lowest of the Four Spirit Worlds called Asiya/Action—it is through action that sexuality performs; driven by the emotions of the heart and the chaotic stream of thoughts from the mind sexuality ultimately must be acted out. The action of sexuality—in and out—is a replication of the first movement of creation called **Nuti Valo Nuti/Inclining and Not Inclining** a movement that describes the entirety of creation as entering and receiving.

The common fascination with sexuality and its relationship with Will and Pleasure make it possible to experience sexual bliss in every part of the body from the ethereal thoughts of the brain to the vibrating sensitivity of the toes. Sexuality encourages the heat of the flesh to raise, the heart to pound and the mind to rage as the entire body takes on the primal in-out motion to coax forth that which is buried deepest inside to come out. The need to periodically participate in this dance of life defines our species more than any other activity.

Animals have appointed times to mate and that is that, but human females menstruate and ovulate monthly and have the capacity and the God given right to chose whether or not to become impregnated for forty days after conception before the law of the embryo falls upon the fetus. There is also another difference between human and animal, the animal always mates from behind whereas human beings can also have sex face to face.

The nature to mate from behind comes from the world of Tohu from where the animal emanates; in Tohu the Ten Sefirot lined up one behind the other growing from the triangle of the head in the form of the snake. The human form is based on three pillars: left, right and center; the center pillar is composed of knowing in the brain, heart in the body and sex between the legs. Our ability to mate face to face amplifies the pleasure and indicates a deeper bond— gazing into eyes and kissing upon the lips.

The mystery of sexuality can be understood through the metaphor which Moshiach represents and how the Earth is a large body curled together like an embryo ready to be born and come awake. At the conjunction of the fourth and fifth millennium while the smoke of destruction was still fresh and the Diaspora was ever evident that the stroke of sexuality, the energy of going in, **Netzach/Victory**, turned to go out into the millennium of **Hod/Retreat**. Many people felt at that juncture Moshiach was coming, but it did not happen and the people went off into exile.

COMING

There is a pervasive feeling within humanity, pervasive as sexuality, which stir the emotions of the human spirit and giving faith that the future is culminating before us. Each human movement upon the planet along with all animated life pushes towards this promise of fulfillment because it is implicit in the process of creation, but just as in the human body, there are places more or less sensual and there are other places which are plainly sexual like the hips and the breasts and the lips, so too does the earth have its fecund places, lush valleys and high peaks.

By the end of the thousand years called **Netzach/Victory** the Jewish people were ready, they could feel the coming of Moshiach, but Moshiach did not come and the Jewish People began to leave the place where God had planted them; it is then that the Jewish People began to pray a prayer that would continue for 2000 years beseeching God to let Moshiach come. That which had been pleasurable suddenly became a burden. The weak ones lost faith and became impotent.

The coming of Moshiach is the orgasm of creation. No one knows what movement, emotion or thought will precipitate the coming of Moshiach and although we yearn for this coming tasting the sweet nectar still our anticipation of the future is clouded in mystery: how will the world suddenly become peaceful? All that is known about Moshiach is that he will bring a new language of clarity and then everyone will understand.

The Tower of Dovid from where Moshiach will come is in opposition to the Tower of Babel that broke language and through that brought about much destruction because of misunderstanding; Moshiach will bring a new, visual language and there will be the clarity that comes with a common vision. Moshiach is that visionary who points the entire world in the right direction not through teachings or books, but by his actions and his ability to communicate to the world.

Compared to Moshiach there is a dark shroud that surrounds the Earth; the coming of Moshiach will bring about a new era of light and peace not promises and extensions. Moshiach incarnates and exists in every generation but has yet to come. There is a certain P leasure that we can do for God that will precipitate the coming of Moshiach, but we do not know what that is; we fervently pray and study, we perform the 613 commandments, but still Moshiach does not come.

As it is in the sexual act, the climax comes suddenly, unexpectedly brought about by a sometimes spurious act anywhere in the body, so it is in our world whose history has largely been focused on the little strip of land at the end of a sea with a narrow opening to the ocean; it is on this little strip of land where three peoples were conceived and born: Arab, Roman and Jewish. The succulent Land of Yisroel is the verdant place of woman's charms; she is **Malchut/ Royalty** where Heaven and Earth merge.

KING

Although the tenth of the Ten Sefirot, **Malchut/Royalty**, is always portrayed as male-king nonetheless this last level known patently as woman is both male and female. The dual nature of this concluding Sefirot indicates the revelation of a secret; speech is related to royalty since the king rules by his word. The king exemplifies ultimate power, but it is the queen who really speaks because words are emitted from the lips and the lips are indicative of woman.

Malchut/Royalty is a manifestation of essence called by the cabalists: **Keter/Crown**. The crown is formed from the merging of two opposites into one—like metal being stretched into a circle. The circle of the crown is represented in the oval of the lips associated with woman. The dual nature of royalty—king and queen—as it says in the Midrash-Book of Metaphor: a king without a queen is not royalty nor is he great.

The two parts of Keter/Crown are Will and Pleasure which is why sexuality as portrayed in the ninth of the Ten Sefirot enters into the lips of **Malchut/Royalty** and produces speech.

Moshiach is a master of languages and he is king—king over the earth. There are two aspects of king; there is the external king who applies the power of will through the statue of his position and there is the internal king who rules through the wisdom of his words taken from the fountain of Pleasure drawn from the pool Truth.

Moshiach is destined to fight wars but has no army, Moshiach rules the world but not through force. This climax in human history upon which every deed is predicated is a premonition in the heart of every human being. What could be more needed then a righteous king to lead the world? It says in the Zohar-Book of Secret there is no king without a people; a king is only a king if the people put the crown on his head. This symbiotic relationship between the king and the people is inexplicably linked to love.

It is the love for the people that keeps the king connected to his subjects, yet it is his aloofness which defines him as king; the importance of this relationship is in the metaphor it represents to each individual person who has the capacity to speak and through his words rule their world. The king-subject relationship has subtle teaching related to the relationship of man and woman, and how to care for offspring. The king is the living example of how man is supposes to rule life.

In our time the monarchy is powerless, usurped by the power of politics where there is only self-love. A pillar of stability was removed with the king and the power of love became like a beacon searching the dark sky for something to focus upon. Where did love go? The preoccupation with the external workings of the world without giving acknowledgement to story of creation wherein the human predicament is encased is like dancing the pig; it is the pig whose **Kosher/Fitting** appearance is but an external presentation— within is vacuous and ill-spirited. The external is like a shell, a mirror where everywhere you look you only see yourself.

Moshiach will break that shell and reveal the truth and because of that he will be king.

OIL

The word Moshiach derives from the word Mesheck the term used for pouring oil. Oil takes greats prominence with the Jewish People along with grain and wine. Grain is substance that feeds the body while wine feds the soul as is says: enter wine exits **Sod/Secret**. Oil is related to fire which produces light—an illumination beyond body and soul. The production of food from grain requires husking, grinding, sifting, mixing and baking; grapes require time to ferment and then there is the separating liquid from drags; but, oil is merely pressed and as soon as the drop comes out it is ready to be used.

Oil has two peculiar qualities: it floats upon liquid and oil ignites when touched by a flame. Floating upon liquid indicates a higher level of existence; Torah which is compared to water is covered by the oil which brings about illumination. Torah, like water, had the ability to descend from a very high place yet not lose or change the simplicity of being; oil is embedded into every substance and can be squeezed from its hidden recesses into the open air through the ether of water.

All three categories are necessary for life: the substance of the body, the energy of the soul and the expectation for the future as symbolized by the nature of fire. In the Temple the oil was administered by the Kohan Gadol who would light the seven wicks dipped in oil and also mix the oil with incense to anoint a king. Where other kings were crowned by the Will of the populous, the Jewish People, who were commanded to make themselves a king as one of the 613 commandments, do so by pouring pure olive oil on the head of the king then drawn down over the forehead in the place of the third eye.

The book of Psalms written by Dovid, king of Israel, says speaking about Moshe's brother and first Kohan Gadol that, the good oil is on the head and beard of Aaron. Aaron exemplifies kindness and love while the oil on his head and beard is about Moshiach who will come with a kindness so great that Avraham who replaced the angel of kindness while on the Earth will have to remove from the right side of God and sit on the left so Moshiach can take his place. This analogy indicates the degree Moshiach will shake the world awake with kindness.

In the Zohar-Book of Secret many dates are given for the arrival of Moshiach, called .; the last one was 5700 or 1940 when the son of the man who wrote the book 666 was saved from the Nazi Holocaust on the last boat out of Poland and arrived in the United States like a drop of oil being squeezed from the olive.

The soul of the Jewish People were being decimated, her meager numbers cut in half, but still the thin delicate line that led back thousands of years, the unbroken promise of the Creator to send Moshiach, still wells within the hopeful heart of every human being.

BAAL SHEM TOV

Thirty years before the American Revolution of 1777 a Jewish Mystic in Poland whose name was Yisroel wrote a letter to his brother-in-law living in the Land of Yisroel then called Palestine, since the time the Roman king Hadriaan 1900 years ago and also by the ruling Turks.

This is his only written record. It is said: those who believe all the stories attributed to this man are fools, but those who believe none are greater fools.

During Rosh HaShana (the celebration of the New Year) in the year **5507/1747** I performed the seven elevations of the soul, as it is known to you. And with awe I saw great wonders—what I had not seen from the day I stood upon my own ability; what I saw and learned in my ascension is impossible for me to relate even if I spoke to you face to face.

But, when I returned to the lower Gan Aden I saw many souls both alive and dead who knew me and knew me not in numbers without limit and they were busily going to and from one world to another by method of the **Amud/Monolith**. The knowing of what I knew with great joy is beyond the mouth to utter or the ear to hear. Also, many wicked people who had repented since it was Rosh HaShana/New Year their misdeeds were repudiated.

Since it was a time of grandeur and even I was amazed who were received back—you even know some of them. They were very happy also because they had also ascended as I had and together persuaded me and pushed me until **Bush/Embarrassment**—the mouth of Rome. Having the Torah and esoteric secrets and a special understanding to grasp and to know these ideas, they invited me to go with them and help them.

Because I saw the happiness within them I decided to ascend with them. I asked my Rebbe (Eliohu) to go with me since it was a great danger to ascend to the higher worlds. From the time I stood on my Amud/Monolith I never went so high—one level after another until I arrived at the abode of Moshiach. There Moshiach was teaching Torah with the people of the Talmud and other **Tzadikim/ Righteous**; with them with the Seven **Raiyim/Shepherds:** Avraham, Yitzchok, Yaakov, Moshe, Aaron. Yoseph and Dovid

I saw they were all very happy, so happy I could not understand why they were so happy. I considered perhaps I was dead, God forbid, but they assured me I had not yet died being with them above while I was making **Y'chudim/Unities** below according to the Torah— but the reason for their happiness until this day is unknown to me.

I asked Moshiach: when will you come; he answered me thusly: in the time when your teachings are known, revealed in the world as springs gushing outward. What I will teach you and the understanding will enable also those to do these **Yichudim/Unities** and **Ayliot/ Ascending** as you.

Known as the Baal Shem Tov, he began a dynasty of learning based on the teachings of Moshiach of which we are in the tenth generation. The son of the author to the book 666 arrived in America 5700 or 1940 the last **Katz/Ending** mentioned in the Zohar-Book of Secret. He brought with him this learning and planted it in America the seed which would grow until fruition. For ten years he taught and then passed away leaving behind him the books, teachings and a modest school.

TIME

The grandfather to the man who wrote the book 666 once had a dream. In the dream he was walking with his grandfather and two other men. One man had a staff and a red beard; the other man was tall and walked with a limp. Along the path where they walked were a stream; the man with a red beard stopped and put out his staff over the water of the river. The tall man with a limp walked out over the calm water and after his return the group continued their walk. At a place in the river where the waters were rapid the man with the red beard put his staff over the water and his grandfather walked out upon the staff and returned. The dream was over.

In the morning the grandfather's grandfather informed him that the man with the red beard was the Baal Shem Tov and the man with the limp was his student known as the **Magid/ Teacher**. The Magid, after the Baal Shem Tov passed away, set up a school devoted to the teachings of his Rebbe the Baal Shem Tov, but he only taught the wise. The grandfather of the grandfather to the man who wrote the book 666 had been the youngest student at the Magid's school. It was his duty to go out to the tumultuous waters of the world, bring the secrets to the nations, awakening the yearning heart so Moshiach could come.

The Jewish laws concerning human sexuality are different from all the other laws which strive to teach how to perform an action, but laws relating to sexuality are the opposite extending to all parts of the circumstances while having no bearing on the actual act itself. Similarly, the one act that will bring Moshiach is unknown and therefore all human actions are equal; the laws and prophecies about Moshiach also have to do with the circumstances around the coming but the people must remain unrestrained to bring about the coming of Moshiach.

As in any sexual union things can intercede and break the mood—ephemeral feelings suddenly reduced to flesh and exploitation follows. The idea of Moshiach is the most luxurious of kind thoughts and mental meanderings; to entertain such an idea requires gentle sensitivities and childish innocence. It is like a young person's first experience with the opposite sex—if it is done with sensitivity and innocence then what follows will be respectful and beautiful, most likely.

The Baal Shem Tov once took the Magid on a trip to a river in early spring when the ice was breaking apart. From a hillside they watched as the villagers cut crosses out of the ice then marching through the streets burned every Jewish house. "See," said the teacher, "what coldness brings upon the world."

After two thousand years of Tohu followed by two thousand years of Torah culminating with two thousand years of Rome the world is ready for Moshiach and Moshiach is ready for the world.

Chapter Ten

Speech

CONCLUSION

Among the stranger stories told in the Torah is the tale of Korach and his assembly who during the forty years wandering in the Wilderness strove to usurp the place of Moshe by demanding an opportunity to be Kohan Gadol: administrator between Earth and Heaven. It was the distinct honor of the Kohan Gadol to light the Golden Candelabra and burn the incense.

Korach challenged Moshe for the right to be Kohan Gadol making the position a populous appointment instead of a directive from God; Korach reminded the Jewish People of their essential being which was revealed when the Torah was given equally to all the people, therefore Korach reasoned it was not right that Moshe should appoint himself leader and his brother Aaron Kohan Gadol.

Along with Korach were **Datan/Law** and **Avirom/My High Father**. The word Korach means Ice and is also the word to describe a practice of those serving other gods who would carve into the hair of their head or flesh of their skull to become a **Galuch(**shaved head)/

Priest. These four names indicate the substance of their rebellion fixated on the cold law of the high father who is in heaven—blaspheme. Dat as in **Datan/Law** later became the word to describe religion.

In the end Moshe promises that God will create a special kind of death; the ground opens up and swallows them all plus the 250 people of their assembly—they are heard screaming from beneath the earth. The rebellion of Korach had been even worse than the building of the Golden Calf when the Jewish People were like children not knowing which way to go. Korach started his rebellion within the very nucleus of the Jewish People—the **Kohan/Server of God**. Korach did not understand that Aaron was Kohan because that is what God wanted, he thought Moshe was just taking power; Korach did not want the power of the Tribe of Levi to be centralized in the family of Moshe—after all they had all received the Torah equally. This story is the epicenter of modern day religion—a thing which requires a new type of death and to be buried deep within the earth from where it will never regenerate.

After the rebellion was swallowed up by the Earth who opened her mouth beneath Korach and his gang, God commanded Moshe to place a staff from each of the twelve tribes of Yisroel in a circle and put Aaron's staff in the middle; in the morning Aaron's staff was budding with **Shakadim/Almonds**—a sign of Aaron's preference by God who is beyond nature. The Midrash-Book of Metaphor says the staff sprouted almonds because they are the first of the tree to blossom in the land of Yisroel where their ivory glow in the hillside is a sure sign spring has begun.

The inheritance of Korach was the residue of his ideas which remained in the psyche of the Jewish People for centuries until the chasm that Korach had opened gave birth to the Tzadoki. The Tzadoki challenged the veracity of the Rabbis by casting doubt on the Oral Tradition: the Talmud, Midrash and Zohar—Books of Law, Metaphor and Secret. The Tzadoki eroded the faith of the people and turned the common dialogue into political banter that they had learned from the Greeks and Romans.

The argument of Korach was convincing in its logic, but life is much more than logical particularly when speaking of inculcating God into reality. The Tzadoki went half way by accepting the Written Law, but refuting the Oral Tradition was like living life without a soul. If the Jewish People lived as the rest of the world lived, they would be giving up their Jewish Soul. Without the Jewish Soul, so says the Talmud-Book of Law, the world could not exist like the Earth can not exist without the wind.

It is the soul of the Jewish People which conveys a light from beyond creation, without that light the human being would be bereft of an essential element; life would lose track of reality becoming obsessed with zero and infinity; without the essential light that radiated through the broad perimeter of the Temple's windows the world would be reduced to rubble. The Dark Ages would soon engulf the world. The sun over the Jewish People had set, but in the sky was illuminated with the splendorous array from the settling residue of the light. Sunset is a point of transition, a declination between day and night that somehow transfers one state to another through what is mysteriously written in the skies as the sailors say: red at the morn sailors take warn.

The concluding destruction of Yisroel ended in 3880 years from the creation of the first human being and 120 years before concluding the thousand years of Netzach/Victory. The Rabbis had won permission from the Roman government to gather in the city of Yavna for the purpose of committing the Oral Tradition into a written text. The Oral Tradition went back two thousand years to Avraham and Sara and continued through their children and grandchild until they went down, seventy in number, into Mitzraim to become a nation.

It was not the history of the nation which was of importance, but rather significant events that reflected on what was later written down in the Torah. Each letter in the Torah is representative of a Jewish Soul and the accumulated 600,000 letters is analogous to the body of the Jewish People whose acts are contained in the Oral Tradition. What had been learned throughout the 2000 years from Avraham to the destruction of the Temple was to be written down in three levels: law, metaphor and secret. Without these three amenities the Written Torah would make no sense— dictates without explanations or meaning.

The time of the Oral tradition was over and it needed to be written down and recorded because the Temple was destroyed, the rebellion crushed and the people cruelly expelled from their land; without writing down the Oral Tradition all would be lost. The Rabbis saw what was happening and decreed the time had come to commit the Oral Tradition into writing named Mishna/Teaching.

Torah which also is translated as Teaching comes from the root of **Ore/Light** whereas Mishna is built upon a much more complex structure like the prism from which light refracts. The word Mishna literally means: repetition, second, and learning; the word Shana, the root of the Mishna, means: change, time and old.

The fascinating story of how the Oral Tradition became a book and how the book progressed through two thousand years of exile has its root in the sunset of the Jewish People and continuing until today when the impenetrable darkness is turning grey. But, beyond the story are the segments of time—the thousand year periods—that alert us to the intrinsic metaphor wherein lays the secret.

The entire Torah, meaning the Written Word and the three levels of Oral Tradition, is compared to water that comes from a very high place then descends forever, never losing her essential self; like water distilling through the Earth seeking the sea whose mighty oceans plume the depths of the Earth longing for her fiery orb, so too the ocean of Torah which the Creator glimpsed—the reflection of God upon the water of Torah—is the light that pervades creation.

The manner and fashion by which the light makes way through creation is by constant contraction like muscles contracting with strength. The process of attenuation is a mathematical construct which can be studied through empirical observation; illuminated by the Torah, physical reality renders the secrets of creation. The exile is the water peculating through the Earth leading to the sea and the light of morning upon the horizon is the secret, beginning to be revealed.

The plan the Creator had put before the Jewish People was riddled with difficulty and pitted with danger, but as one accepts the brightness of the day so too the darkness of the night must also be accepted. The Jewish People were taken from the Land of Israel because the world was prepared to receive Jewish knowledge, but also because it was a way to purify the people from their iniquities. The Talmud-Book of Law clearly states: the Temple was destroyed and the people removed from the land and sent into exile because of **Sinat Chinom/Free Hatred**.

Anger, as described in the Torah, is like serving idols; to be truly angry necessitates separation from God and a rejection of the creation. It was from Korach and his gang who exemplified this anger and revealed it through hatred without a cause—just pure unadulterated hatred for no reason—**Sinat Chinom/Free Hatred**. Hatred begets hatred and soon the people hated those who brought hatred to the people, but before it could be eradicated the people began to hate one another.

Hatred became enshrined within a movement called the Tzadoki dedicated to the destruction of the Jewish People as defined in the Torah. Like a remnant of Haaman who tried to physically destroy the Jewish People, the Tzadoki were prepared to kill the Jewish Soul. The apparatus of the Tzadoki exists even today stirring the pot of discontent. Because hatred is as blind as is love.

It was because of self-hatred the Jewish people were sent into exile to be strangers once again in a strange land—in many strange lands; each land represented a different part of the human being upon the Earth where the Torah like oil quietly penetrated and then moved on. As the Jewish People became closer to God through the distillation of knowledge and the inter- activity of people with their cultural interchange, the Tzadoki sought refuge at the hand of the enemy.

The Tzadoki could not go beyond zero and infinity; they were weak and lacked the faith to go beyond what they understood. Confined to logic, the Tzadoki formulated ideas and proposed solutions that did not require Godly intervention. God was looked upon with suspicion by the Tzadoki who, like Korach and his gang, did not understand how the line becomes a circle. Instead, they saw the line and the circle as two independent entities that exhibited opposite qualities—nothing and infinity expressed in the modern day as zero and one.

The Tzadoki are a disease of the essence as is explained in the book 666: the dual nature of essence composed of male and female qualities derived from the dot and the line produced from the previous worlds of **Akudim/Lines** and **Nikudim/Dots.** The external part of essence is the male line like a beam of light which comes from and is received by the dot. To the Tzadoki the dot served to segment the line from the Heaven to the Earth, but the Rabbis knew and taught something different.

According to the Rabbis, life is not a line but a circle where God connects to the heavens made from spirituality that enlivens the Earth and then the Earth connects directly to God; the Tzadoki, like the scientist, leave God out of the equation. It is furthered explained in the book 666 that when the essential will is performed here on the physical Earth thus finding fulfillment connects back to the Hleasure taken by the Creator in the workings of creation.

At the time of the destruction of the Second Temple, Rome also was dividing into two; the royal line fathered by Esau had ended with Julius Caesar, the last ancient pontiff of Rome which led to the ascension of Vespasian to the throne. Vespasian was from the common people whereas the royal families of Rome were from the land known as Pontus rendered in Aramaic as Shenar the place where the Tower of Babel was built—Targum Yerushalmi. The Pontiff with his unearthly hidden powers was equivalent to the king.

The Tzadoki, having more in common with the descendants of Esau than the commoners of Rome, conspired with the Pontiff to bring down Rome and take all power for themselves. After the destruction of the Temple and the crushing defeat of the rebellion of Bar Chochba the Tzadoki disappeared—they were helping the new pontiff resurrect religion from where it had been buried beneath the Earth in some hellish unimaginable place—the people who witness the Earth swallowing Korach along with Datum and Avirom heard them screaming from beneath the ground. **Dat/Law,** like in Datum, is used in modern day Hebrew to convey the concept of religion and dogma.

Since the time that Esau sold his birth right to his twin brother Yaakov and was forced to leave the Land of Yisroel, he went to the place of the Etruscans on the land known today as Italy and began a nation built on the proverb that one brother kills another. When

Rome became a consolidated power, Alexander the Great had already conquered many lands taking back to Greece not only the booty of gold but also a wealth of knowledge accumulated from various nations. When Rome defeated Greece they inherited this knowledge and the gods they had erected.

For a few centuries Rome merely shaved and clothed the Greek gods giving them new names and appropriating their previous roles, but in order to claim the future they would have to devise something original and that is where the Tzadoki came in. The Tzadoki wanted revenge from the Rabbis whose severity in rulings cause the destruction of the Temple; the Oral Tradition which the Rabbis were undertaking to compose, the Tzadoki wanted to erase.

By substituting a new idea, the Tzadoki could ingratiate themselves with Rome while destroying the Rabbis and what the rabbis sought to restore. The strange story of Jesus, the disputed last prophet of the Jewish People, served just that purpose by being enigmatic and miraculous. The Rabbis had called him a **Momzer/Bastard** and had refused him membership into their ranks. He had died and regenerated a new body; the Tzadoki were in possession of scrolls, some claiming to be eye-witnesses; they took what would support their fabricated story and destroyed the rest.

Vespasian had done the same thing, using the history of the Jewish People as written by Josephus while eliminating all other accounts that conflicted with his account. In the chasm between the Pontus and the people was born the collusion between government and religion; incubated in Rome then spread throughout the world.

The Religion of Rome, the Government of Rome, the Coin of Rome, the Technology of Rome, the Sword of Rome and the Roads of the world all lead to Rome are like tentacles around the Earth inculcating their brand of god.

There is nothing more powerful than the idea of God—there is not a people who are not cowered by this profound concept—Creator of Heaven and Earth; there is not a child born who does not wonder where everything comes from and where it all is going. The question belies natural humility and it is through humility that we come closer to God. Such is the fear and awe of God imbedded within each human heart that those who come in the Name of God have an advantage— sent by the Almighty who will assure victory.

Rome comes with his sword and kills those who will not accept their religion then governs those who remain prodding them with money and technology that ties them to the empire; quickly the Roman language and culture follow until Rome's conquest of the world is complete—even then Rome does not promise peace. Because Rome is stuck between zero and infinity war is inevitable. Rome is very strong on the outside but weak within, lacking faith and conviction they invent dogma delivered with severity.

But, the most oppressive of Roman ways has been the calendar—abrogating everything that went before and reestablishing time according to the birth of their god. This practice continues unabated until today. The date they refer to is the birth of Jesus, a Jewish prophet who came to make peace between the Tzadoki and the Rabbis, but instead was killed by the Rabbis for being a false prophet and exploited by the Tzadoki in their coalitions with Rome. Rav Tzadok fasted 40 years from the time Jesus went public until the destruction of the Second Temple.

The perfidy that transpired over this one man is enormous, but it occurred because it is the way to truth—as the Rambam a thousand years later writes, concerning the laws of Moshiach: Yeshea HaNotzri (Jesus) made himself to be Moshiach and the **Beit Din/Rabbinic Court** put him to death. As prophesied in Daniel—the son of a harlot of the people will raise, erect a vision and stumble. Is there a greater stumbling block than this? All the previous prophets spoke of how Moshiach would deliver and save Yisroel, gather together the scattered, strengthen **Mitzvot/Commandments**, but this one cause the destruction of Yisroel by sword, scatter the people to the wind, changing the Torah and fooled many people in the world into thinking that God is something other.

Time is a spiral wide and broad at it conception, but narrow and hurried at its conclusion—a light bulb about to extinguish. But, what is light to us is water to God so the conclusion of this last six thousand years of male energy is to spread out like water beneath the sun and rest for a thousand years in the lush arms of woman. It is the distortion of time that acts as a veil around the Earth blocking the sunshine and casting a long shadow. The Golden Candelabra from the Temple carried into Rome remains until today in some dark cavern—a light was put out and shroud lay upon the sky. In truth, it was just the transition from one millennium to another. The attribute of

Netzach/Victory had run his course; symbolized by the right leg he was soon overcome by the left.

The sunset of the Jewish People was accompanied by the transition between victory and retreat beneath the red sky. The die had been cast for the next two thousand year segment divided between **Hod/Splendor** and **Sod/Secret**. The energy of **Hod/Splendor** is the opposite of **Netzach/Victory**—one goes out and one goes in. After the destruction of the Temple the Jewish People physically went out, but spiritually went in, built small communities throughout the world and waited for the time to return back to the land of Avraham, Yitzchak and Yaakov where the 613 Commandments can be performed and the Temple can be rebuilt for a third time.

It is said in the Midrash-Book of Metaphor the Third Temple will last forever because it will descend from the heavens made of material that does not degrade, yet it is a **Mitzvah/Commandmen**t for man to rebuild the Temple—this is the impetus that will bring the Temple forth from the heavens.

The Book 666 explains how the final action connects to what preceded the beginning by questioning why God's light is termed OreAinSof/LightWithoutEnd instead of **OreAinTichila/LightWithoutBeginning**? Beyond beginning is much more mysterious than forever. In his answer is explained three states of essence: the hidden light of essence, the essence and the point in between when essence begins the journey into revelation. When the final action conceived in the essence and perpetrated into Will is received and accomplished the resultant pleasure eliciting new pathos within the Creator, Moshiach will come.

The essential sexual metaphor implicit within the story of the Jewish People and exposed through the secret of the Torah is the key to unlocking time and foreseeing the coming climax. In the beginning time was wide and luxurious but now time is rushing and the world is awake because something is going to happen—we feel it coming in waves from somewhere deep within.

Those who believe life exists only between zero and infinity construct fabricated constraints to try and wall out reality, but those who see the pattern of creation permeating in the most unlikely places, geometric living form, know it is not chaos that surrounds us but endless light of perfect portions in constant motion. The purpose of creation is to manifest the original desire of the Creator from before there was a beginning; as it is in our physical reality the space between the Earth and the Sun conveys the light so too the heavens convey the light of God—but it is upon the earth that the deed must be done.

Speaking in sexual terms about interacting with the Creator helps us to understand the rabbinic law to be naked when copulating because nothing is hidden when you are naked. Similarly, in our relationship with God it is required to remove all garments particularly the filthy ones. Unfortunately, we are all mired in misdeeds and rebellion.

Strangely, the most contentious issue—Jesus, is the key to Moshiach since the root problem is the friction and disparity between the Tzadoki and the Rabbis; what has transpired in the last 2000 years has been a direct result of that gap. Religion and scientific secularism propped up by governmental decree are the modern representations of two similar but opposite points of views: the secular perception of creation between zero and infinity and the religious view point of instantaneous manifestation. Both sides skew the facts. Science refuses to incorporate into their theories anything other than extrapolated conclusions based on measured entities which exists between zero and infinity whereas religion sees creation as a crucible of trouble so chaotic and without explanation that only dogma can redeem from its fiery wrath. Of course, neither is true, but by skewing the facts of reality each is able to propagate a philosophy based on logic and morals derived from belief.

Because the premise is faulty all that follows is on shaky ground. Morality of religion does not reflect justice and philosophy built on logic does not prevent war; in fact, it is religion that promotes war through sectarian violence and pointless hatred. Faulty history which emphases the chronicling of war and religious; inaccuracies as to the nature of the creation has led man astray. Only the truth will set us free.

One of the major conundrums with the story of Jesus is the complete lack of evidence even though great historians lived during that time nothing is said or referred to that corroborates the existence of Jesus. There is no record of Jesus or any such proceedings in the courts of Pontus Pilate the governor of Judea during that time—no evidence of such a person being crucified. How it is that such an undocumented story accusing the Jewish People of having given one of their own over to the Roman Empire to be crucified could have pervaded all walks of life throughout the world?

Because Rome wants to rule the world, the concept of Moshiach became known everywhere albeit through a fabricated false story that indicted the Jewish People. They stripped the Torah naked and discarded her garments: the three levels of the Oral Tradition. While the Torah was dragged away into Rome, naked and ashamed, the Rabbis assiduously worked those discarded garments—The Temple was gone and all that was left were the garments. As a parting shot, the Tzadoki changed one letter in the Torah in the copy that they took to Rome— insignificant to Rome since the letter changed did not alter the meaning of word. In the fight between the Rabbis and the Tzadoki, the Tzadoki were going for the balls—like Amalek attacking sexuality.

But, for the Jewish People it was much more than being attacked at our most vulnerable place; any change to the Written Torah was strictly forbidden and carefully monitored; only by maintaining a strict scrutiny over each letter could the veracity of the Torah as handed down from Moshe be preserved. This breach at the very core of the written text represented a huge problem for the Jewish People migrating throughout the world.

The number of letters in the Torah is approximately the same number as the people who received the Torah and entered into the Land of Yisroel to be divided up according to families—a sign that each person has their place as it is explained in the laws conducting the writing of the Written Torah: if one letter is disturbed or missing the entirety of the Torah becomes invalid. By altering a letter, the people in the north would always be divided from the people in the south.

Jesus came with a very simple message delivered not in words or pictures but in action— what happened to him after he died because this was the fulcrum of discussion between the Tzadoki and the Rabbis. The Rabbis insisted that it is better there should be no reward because then one's actions would be completely for the purpose of God and then perhaps Moshiach would come; the Tzadoki maintained, perhaps there was no reward and therefore why not become like other nations.

Jesus, the last Jewish prophet, was the first to undergo the physical transformation promised by the previous prophets: the growing of a new body from a bone buried in the Earth; this new body would be a vehicle to encompass the whole of the soul and people again will live 1000 years. The message was: not only is reward for the soul but also for body. But, there was a deeper component that went back further—to the incident of Korach 1500 years before Jesus came to deliver his message.

The only Hell recorded in the Torah is the one prepared for Korach, Datan, Avirom and their followers when the earth opened her mouth and swallowed them live. These three names translated together produces: Cold Male Religion—the original seed of the Tzadoki. How all of this has been played out over the last 2000 years is the subject of the fifth and sixth concluding books of The Dance of the Pig.